INTO STABILITY

WALTER R. EVANS
and the
**STORY OF
ROOT LOCUS**

GREGORY W. EVANS

Foreword by
ROBERT H. CANNON JR.

Into Stability: Walter R. Evans and the Story of Root Locus

By Gregory W. Evans

Copyright © 2025 Gregory W. Evans

All rights reserved. No part of this book may be reproduced, distributed, or transmitted in any form or by any means, including photocopying, recording, or other electronic or mechanical methods, without the prior written permission of the publisher, except in the case of brief quotations embodied in critical reviews and certain other noncommercial uses permitted by copyright law.

Published by Evans Heritage Press, Los Altos, California

ISBN: 979-8-9931614-9-5 (Paperback B+W)
979-8-9931614-8-8 (Hardback – Color)
979-8-9931614-0-2 (Hardback – B+W)
979-8-9931614-7-1 (Ebook)

Library of Congress Number: 2025921847

First Edition

Printed in the United States of America

For permission requests, contact:

Evans Heritage Press
greg@walterrevans.com

DEDICATION

*Dedicated to the Memories
of the Men and Women
Who Helped Defend
the Free World
During the
Cold War*

Advance Praise for *Into Stability*

Richard Murray
Professor of Control and Dynamical Systems, Caltech

"Greg Evans tells the fascinating story of his father, Walter Evans, whose development of the root locus method transformed control engineering. From foundational ideas to the clever 'spirule' tool, this book captures the blend of insight and ingenuity that shaped the field. An engaging read for anyone interested in the history of engineering innovation."

Shalom Ruben
Teaching Professor of Mechanical Engineering, University of Colorado, Boulder

"The son of Walter Evans weaves a compelling history of a 1948 transformation in control engineering that remains a cornerstone of engineering education and practice today. This is a story of how family, teachers, and colleagues created an environment that nurtured the critical thinking and resilience needed to bring such an out-of-the-box idea to the world."

Qimin Yang
Professor of Engineering, Harvey Mudd College

"This book provides an intimate memory of Evans's life and his invention, his deep connections to people in academia and industry, and the critical impact of his work in the postwar aerospace industry and training of the control engineers. I highly recommend this inspiring story to any engineering students, particularly those who are interested in controls."

John Doyle
Professor of Control and Dynamical Systems, Caltech

"This is a wonderful book about a giant of classical control who inspired me as an undergrad at MIT. I devoted my early career to bringing back the rigor and relevance that Evans and his contemporaries embodied and was sadly missing in modern control. Reading this book reminds me of what I missed in knowing the methods but not the man."

Don Evans
Former US Secretary of Commerce

"Fueled by a passion for lifelong learning, teaching, and love of country, Uncle Walter was a humble and brilliant soul, and one of America's most impactful engineers. After a debilitating stroke at sixty, with the love and support of Aunt Arline, he taught himself to paint left-handed, continuing to grow, create, and inspire with every stroke of the brush."

David A. Peters
McDonnell Douglas Professor of Engineering, Washington University in St. Louis

"This is the fascinating story of Walt Evans, the man who invented the Root Locus design tool—and changed the world. We would not have made it to the moon without this methodology, which is still in common use today. This book is a must-read for anyone interested in the history of technology and in the human stories behind it."

Walter R. Evans c. 1950

Foreword

by Robert H. Cannon Jr.

Two things came together in the late 1940s: The remarkable young people in the auto-navigator division of North American Aviation's Aerophysics Laboratory and one of its new leaders, Walter Evans. This team was solving very difficult engineering problems one after another, to produce, for the first time anywhere, precise navigation systems for very long-range, unmanned aircraft and for submarines that went exactly to the North Pole, among many firsts. These were very hard systems to achieve.[1]

Control systems had to remain stable under conditions and deliver precise performance at speeds and ranges never before attempted. Classical analysis tools—while mathematically sound—provided little intuitive guidance. And perhaps the biggest gap: While it was clear that poles and zeros controlled a system's behavior, there wasn't a reliable way to move them around on the complex plane. That made it hard to design for specific dynamics, especially when stability had to be rock-solid.

Walter Evans, better known to his colleagues as Walt, was a person of remarkable insight and a major leader in the understanding of automatic control and how to design excellent systems very well and very quickly. One of his field-leading contributions was the invention of the root-locus method for seeing instantly the natural dynamic behavior a linear system will have, seeing it directly in terms of the control parameters at the designer's

---

*   A superscripted index at the end of a paragraph references a source in the Bibliography with the same index.

disposal. The method presents—in seconds—a plot of the system's stability, speed of response, and the damping quality of all of its natural motions.[2]

These were very smart young people. They saw at once the power of the root-locus method, and it spread very quickly through their culture at North American Aviation. It was just very easy to learn and use and place at the center of discussion of every control system design. From there, the root-locus method spread quite swiftly throughout the international culture of automatic control. It is still typically part of the first design steps and of the dialog among designers and users.[1]

There are several reasons for that. The roots of a dynamic system's characteristic equation reveal directly and quantitatively the natural behavior it will have—at what frequencies it will vibrate and how quickly the vibrations will damp out. By plotting the locus of these roots versus the design parameter being chosen, one can see precisely which values give good behavior and which will not—which values will make the system unstable, for example.[3]

Before Evans's root-locus method, these natural-behavior design decisions were most often made by the astute but very indirect methods of Nyquist and Bode and Nichols, which inferred them from the system's response to sinusoidal inputs. Good design throughout the years since has used the two methods concurrently. But it is the root locus that gives the first quick, direct insight. And it is the root-locus structure that supports advanced optimal control design methods in a very fundamental way.[4]

Formerly, Evans taught courses at Washington University in St. Louis and at UCLA and presented seminars that always surprised. His book, *Control-System Dynamics*, and his seminal papers recorded succinctly his brilliant conception of automatic control. But the really fortunate students were the colleagues working beside Evans and watching a wonderful, creative, agile, and unfettered mind working with zest and very special wit, which was always brought directly to bear. His goal-oriented mind always approached any problem from a way no one else had ever thought of and often the only way that would work. Every so often, he would make a giant leap. He solved a rocket engine control problem after only an hour's exposure to it. It took quite a while for others to assimilate what his mind had provided.

FOREWORD

Working with Walt was enormously stimulating and full of surprises. In no period in my life did I learn real engineering more rapidly, or more deeply, or with more enjoyment. Walt shared his office with Bill Mullins and I. Bill was also a terrific engineer, and that was a great office for a young, green, fresh-out-of-school guy to be in.[3]

It was in those days that Evans led the development and construction of the stable inertial platforms for the guidance systems for, among others, the Minuteman [intercontinental ballistic missile] and two submarines called Nautilus and the Skate, which used those systems to navigate without any external reference.[3]

Evans saw the new problems coming, and in problem after problem his agile mind made giant leaps over the details to the key to the good answer. A problem for us was that, in his beguiling modesty, he just naturally assumed everybody else's mind worked as fast as his. Bill Mullins and I would scratch our heads and after about an hour we'd figure out what it was he'd just told us in five minutes.[3]

Then, in 1980, tragedy struck. At age sixty, Walter Evans suffered a massive stroke. With his indomitable will and Saint Arline's unstinting support, he continued to bring love and philosophic inspiration to us all, to paint beautifully, and to play chess well and swim often.

He was most appropriately awarded the Rufus Oldenburger Medal from the ASME in 1987, the Richard E. Bellman Control Heritage Award from the American Automatic Control Council (AACC) in 1988, and the Engineering Alumni Achievement Award from Washington University in St. Louis in 1990. And the highest distinction of all was the deep and warm respect of his colleagues: Walter Evans inspired us so much. And he gave us so much.[4]

Above all, Walt Evans was ever devoted to his four children, Randy, Greg, Nancy, and Gary, and to his wonderful wife, Arline, who has been his super-supportive and brilliant companion throughout his life and career.[3]

Robert H. Cannon Jr. (1923–2017) was the Charles Lee Powell Professor of Aeronautics and Astronautics at Stanford University.

Cardinal by Walter R. Evans, Undated, Portfolio #335 Vol. 1

On June 2, 1980, a stroke destroyed 30 percent of the left hemisphere of Walter Evans's brain. He subsequently taught himself to draw with his left hand. His drawings of cardinals honor his hometown, St. Louis, and their baseball team, the Cardinals.

INTO STABILITY

TABLE OF CONTENTS

DEDICATION iii

FOREWORD vii

INTRODUCTION 1

PART I: ROOTS 5

    Chapter 1 St. Louis (1920–1941) 7

    Chapter 2 Schenectady (1941–1946) 17

    Chapter 3 The GE Advanced Course (1941–1945) 26

    Chapter 4 The Graphical Analysis Paper (1946–1948) 38

    Chapter 5 North American Aviation in 1948 46

PART II: FEEDBACK 57

    Chapter 6 Aerophysics Lab (1948–1954) 59

    Chapter 7 The Spirule (1948–1952) 69

    Chapter 8 The Root Locus Paper (1948–1950) 81

    Chapter 9 The Textbook (1949–1954) 91

PART III: STABILITY 107

    Chapter 10 Autonetics (1955–1959) 109

    Chapter 11 The Spirule Company (1952–1980) 120

EPILOGUE 127

APPENDICES 137
    Appendix 1: The Root-Locus Method of Walter R. Evans 138
    Appendix 2: New Challenges for Engineers (1945) 142
    Appendix 3: The General Electric Years (1941–1946) 145
    Appendix 4: The "Root Locus Idea" (1949) 150
    Appendix 5: The Invention of the Spirule (1948–1951) 155
    Appendix 6: Historical Spirule Documents (1948–1951) 159
    Appendix 7: Profiles of Prominent Engineers 167
    Appendix 8: Correspondence from Autonetics Colleagues 171
    Appendix 9: Walter R. Evans Biographical Information 177
    Appendix 10: The Quotable Walter R. Evans 180
    Appendix 11: Pinball: Polynomial Factoring with Root Locus 182
    Appendix 12: The Artwork of Walter R. Evans 188

MY STORIES OF DAD 193

GLOSSARY 209

BIBLIOGRAPHY 211

ACKNOWLEDGEMENTS 214

CONTRIBUTORS 215

ABOUT THE AUTHOR 218

Introduction

The history of control systems and servomechanism design is rooted in the essential human quest to automate and regulate processes. Long before the advent of modern control theory, engineers and inventors grappled with the fundamental challenge of achieving stability and precision in mechanical systems.

The origins of automatic control can be traced to ancient times. Early examples include the float valve used in ancient Greek and Roman aqueducts to regulate water flow. Centuries later, during the Islamic Golden Age, scholars like al-Jazari devised water clocks with feedback mechanisms. These rudimentary systems demonstrated basic principles of feedback control, though the concept itself had yet to be formally articulated.

The Industrial Revolution accelerated the development of mechanical control systems. James Watt's steam engine governor, invented in the late eighteenth century, is often cited as a pivotal innovation. The governor used centrifugal force to regulate engine speed, maintaining a balance between power and stability. However, as industries demanded more precise and responsive systems, engineers encountered increasingly complex stability challenges.

By the early twentieth century, the emergence of electrical engineering introduced servomechanisms—automatic devices that use feedback to achieve desired motion or position—which became central to military and industrial advancements. Systems for gun aiming, aircraft stabilization, and ship navigation required not only mechanical precision but also rapid, reliable control responses. Engineers began to adopt mathematical analysis to predict system behaviors, yet tools for visualizing and designing stable systems were still rudimentary.

As the century progressed, engineering was caught in a race: More complex machines promised new capabilities, but their very complexity bred instability. It was not unusual for costly prototypes to fail or even crash because designers lacked a method to anticipate how feedback loops would behave as conditions changed. The need for clarity became pressing.

In 1948, Walter R. Evans gave engineers a powerful new method to understand and design for system stability: the root-locus method. First sketched on a classroom blackboard in 1948, root locus transformed a field that had relied on formulas and intuition into one where stability and performance could be visualized directly. It became both a practical tool and a teaching language, making its way into textbooks, classrooms, laboratories, and eventually into the software that every engineer uses today.

Blending biography and engineering history, the book follows Evans through a pivotal decade of invention and impact. But Into Stability is about more than a method. It is about the life and character of the man who created it. Walter Evans grew up in a family that prized education and problem-solving, studied under teachers who opened doors to new ways of thinking, and found in Arline Pillisch a partner whose steady support carried him through both triumphs and trials.

His career began in wartime laboratories at General Electric, moved into the booming aerospace industry at North American Aviation, and was shaped by friendships, mentors, and students.

The invention of root locus did not happen in isolation—it grew out of these roots, was refined through feedback, and ultimately proved itself by bringing stability to both systems and lives.

The arc of the book follows these stages:

- **Part I: Roots**   The foundations of Walter Evans's thinking—family heritage, education, and the crucible of early career experiences—combined with the entrepreneurial environment of North American Aviation. These "roots" shaped both the man and the method.

- **Part II: Feedback**   No innovation takes hold without conversation and correction. Between 1944 and 1954, Evans's ideas encountered the influence of colleagues, students, and competitors. The questions of how to teach, refine, and spread root locus circulated in a feedback loop.

- **Part III: Stability** This section reflects on Evans as engineer and father, and on the steady rhythm of his life. It gathers reflections from colleagues, connecting the technical legacy to the personal qualities—humor, humility, integrity—that defined Walter Evans.

- **Epilogue** Tested by a stroke that left him disabled but never defeated, with Arline at his side, he demonstrated resilience and a joy in living, setting an example for young and old, healthy and infirm alike.

- **Appendices** Supporting material, technical notes, firsthand accounts and background documents for those who want to dig deeper into the details of the root-locus method and its developer, Walter Evans.

- **My Stories of Dad** Personal recollections from my perspective as his son. These stories bring forward the father who inspired by example.

Seventy-five years after its first publication in *AIEE Transactions*, root locus remains one of the most widely taught and used tools in control system design. Whether manually plotted with a Spirule or generated instantly by software, it carries forward the clarity and ingenuity of Walter Evans. This book is the record of how it came to be, and of the life that made it possible.

Cardinal by Walter R. Evans, dated January 1991, Portfolio #207 Vol. 1

Part I

ROOTS

Insist on understanding!
Do away with superficiality!
Stop memorizing words and formulas
that you don't understand, merely for a grade.

—Robert E. Doherty, Founder of the General
Electric Advanced Course in Engineering

Daniel Evans (GGF)
Welsh Coal Miner
and Farmer

James X. Allen (GGF)
Union Army Surgeon –
Washington Univ. Grad

Samuel Burgess (GF)
West End Chess Club

Eveline Allen Burgess (GM)
United States Women's
Chess Champion

Gomer Daniel Evans (GF)
Locomotive Engineer

Sybilia Burgess Evans
(Mother) 2nd in HS Class

Evans Family on Farm that Evans's Great-Grandfather,
Daniel Evans (1831–1917), bought in 1872

Chapter 1

St. Louis (1920–1941)

Walter Evans's journey to revolutionizing control systems engineering started in St. Louis, Missouri, a city that had shaped generations of thinkers, innovators, and engineers. His roots, both familial and intellectual, were deeply embedded in this city's educational institutions, its intellectual circles, and its values. Walter's role models and the City of St. Louis were achievers.

A Family That Valued Knowledge

Walter Evans was the youngest of four children in a family that placed a high value on education. His maternal great-grandfather, James X. Allen, served as a surgeon in the Civil War. In the 1870s, James attended a medical college that later formed the foundation for Washington University's Medical School, a remarkable achievement for the time. Walter's maternal grandfather, Samuel Rostron Burgess, was a founder of the West End Chess Club, and his wife, Eveline Allen Burgess, was the United States Women's Chess Champion in 1906. Walter's mother, Sybilia Burgess Evans, was second in her class. Both Walter's father, Gomer Evans, and two brothers, Cedric

Walter Evans in St. Louis circa 1932

and Sam Evans, graduated from Washington University with engineering management degrees, solidifying a family legacy steeped in analytical rigor and disciplined thought. Walter would follow in his brothers' and father's footsteps at Washington University.

Cedric, Walter, Alice, Sybilia, and Samuel Evans in St. Louis circa 1928

Walter's LDS Great-Grandparents

Walter Evans's paternal great-grandfather, Daniel Evans, was a Welsh collier (i.e., coal miner); he was one of four sons born on a farm near the South Wales village LLandyfaelog, a few miles from the port city Llanelli. He left the farm to work in one of the massive coal fields in and around Merthyr Tydfil, where he married Gwenllian Williams in 1853. After they arrived in St. Louis in 1856, he worked in one of the city's numerous clay mines for sixteen years.[6]

Walter's great-grandparents were all baptized in the UK into the Church of Jesus Christ of Latter-day Saints. All eight sailed from Liverpool, England, between 1842 and 1856 on sailing ships chartered by the church. His great-grandparents, grandparents, and parents met at church functions. They then left the LDS church to join what they called the "Reorganization," later to be known as the RLDS church. Walter's parents participated socially but did not accept LDS theology. Three of their four children, including Walter, married outside the church and joined mainline Protestant denominations.

Untimely Death: Daniel's Son, Gomer Daniel Evans

In 1872, Daniel Evans bought sixty-five acres about seventy miles southwest of St. Louis, near Sullivan, and became a farmer. His only son, Gomer

Daniel Evans, was father to five children, including Walter's father, Gomer Louis Evans.

Gomer Daniel Evans became a railroad engineer. He was at the helm of a freight train in 1897 when the roadbed, weakened by storm waters, gave way along a stretch

Evans's boyhood home at 7048 Nashville Avenue

of the Missouri River a mile east of New Haven. All three crew members perished in the river. Gomer's older sons (Gomer Louis's brothers) abandoned their education and went to work to make up for the loss of income. Being the youngest, Gomer Louis was spared. He received a scholarship, earned an engineering administration degree, and became a vice president of Wagner Electric, a St. Louis manufacturing firm.

In 1917, Daniel Evans, the patriarch of the Evans family, died of pneumonia at age eighty-four. The cousins hired Earl Burnett as its tenant farmer. They built a cabin to have a place to stay when family members, including Walter, went to the farm. They worked summers in the field, hunted rabbits, swam at the "swimming hole" on a tributary of the Merrimack River, and rode the farm's horses.

Farm Cabin near Sullivan, Missouri (Franklin County). The family built it as a place for them to stay.

A Second Untimely Death

Gomer Louis Evans (Father)

Tragically, Walter's father Gomer Louis Evans, vice president for Wagner Electric, died due to a surgical accident. At the time of his father's death, Walter was only fourteen years old. Fortunately, Gomer had learned a lesson from the financial impact that the early death of his own father had had on his family. He had taken out a life insurance policy providing his wife a lifetime income. What sort of man was Walter's father? Perhaps H. N. Fifer's poem, "He Lived a Life," recited at his memorial service on September 12, 1934, offers a window into his soul.[7]

> What was his creed?
> I do not know his creed, I only know
> That here below, he walked the common road
> And lifted many a load, lightened the task,
> Brightened the way for others toiling on a weary way;
> This, his only need; I do not know his creed.
>
> What was his creed? I never heard him speak
> Of visions rapturous, of Alpine peak
> Of doctrine, dogma, new or old;
> But this I know, he was forever bold
> To stand alone, to face the challenge of each day,
> And live the truth, so far as he could see—
> The truth that evermore makes free.
>
> His creed? I care not what his creed;
> Enough that never yielded he to greed,
> But serve a brother in his daily need;

> Plucked many a thorn and planted many a flower;
> Glorified the service of each hour;
> Had faith in God, himself, and fellow men;
> Perhaps he never thought in terms of creed;
> I only know he lived a life, indeed.

The Formative Years at Soldan High School

Walter's academic journey formally began at Soldan High School. It was here that he first encountered both intellectual stimulation and companionship.

Geometry was his favorite subject, as evidenced in these comments he wrote to the author's high school trigonometry teacher: "Math has always been a game for me and now is a good part of my livelihood. Geometry used to provide [me] a steady diet of looking for a pattern that would lead to a solution before settling down to the detail of writing down all the steps."

Moreover, in geometry class Walter met Arline Pillisch, marking the beginning of a lifelong partnership. Arline was a brilliant student, becoming class valedictorian in 1937. The two young scholars set out on their respective paths, but their connection would endure. Her steadfast encouragement, patience, and belief in Walter's work were essential in his development. While Walter would go on to make significant contributions to engineering, Arline's role in his life was no less impactful—she was, in many ways, another root of root locus.

Arline Pillisch in St. Louis circa 1936

Two and a half years after his father's death, Walter Evans graduated from Soldan High School. Its yearbook recorded that he served as senior class treasurer. Superintendent of Schools Henry G. Gerling made the following announcement on March 18: "The four year Honor Scholarship to Washington University allotted to each St. Louis Public High School was awarded to Arline Pillisch." Arline, whose GPA of 93.667 over four years put her at the head of her class, would be the first in her family to attend college. Upon learning of his daughter's opportunity to further her education in college, Reinhold Pillisch wept in joy. Walter and Arline would enter Washington University together in 1937, thereby giving their budding relationship an opportunity to grow.

Mentors at Washington University

Rich with intellectual rigor and pioneering engineers, Washington University was where Walter honed his analytical abilities, built critical relationships, and encountered mentors who would shape his thinking for decades.

Roy Glasgow   Alexander Langsdorf   Frank Bubb   Ross Middlemiss

Among those mentors were four professors: Roy Glasgow, Alexander Langsdorf, Frank Bubb, and Ross Middlemiss. These men were not just instructors; they were intellectual guides who introduced Walter to the nuances of engineering analysis, mathematical rigor, and practical problem-solving.

They instilled in him a disciplined approach to thinking—one that would later manifest in his groundbreaking work in control systems. Fortunately, copies of letters he wrote to his professors, decades after graduation, elucidate the impact these men had on him in his own words.

To Roy Glasgow, on the occasion of his retirement in 1966, Walter wrote,

Dear Dean Glasgow. Fond memories provided by you are so numerous that I will have to limit this letter to those comments which triggered some key decisions in my life, or remarks that I have

modified for various occasions. It is hard to believe now that, as a sophomore, I was planning on Engineering Administration. You advised that it would be better to prove myself as an engineer first and worry about the vice-presidency later.

The choice at graduation was between GE and Wagner Electric. You advised me of the glowing comments from alums of GE's Advanced Course but warned about the tough entrance exam. Fortunately, the exam was loaded with your kind of problem [which had] set the hook on my liking to attack any problem to achieve as much of a solution as permitted by the initial conditions of knowledge and the time allowed to respond. The [1947 summer] job you set up at Emerson Electric set the stage for root-locus by requiring a real working-over of the complex plane in trying to get the frequency response cult off their j omega axis.

Walter Evans studying for a GE Advanced Course class, Schenectady, New York, in the autumn or winter of 1941.

INTO STABILITY

In a 1961 letter to Dean Langsdorf, Walter shared what he learned through observation of other students. Throughout his life, Walter had a keen interest in understanding how other people ticked. Here are his observations:

> *At Washington University I found that most students could memorize something like the vector diagram of a synchronous machine ... if the subject was repeated about three times. A basically simpler situation which was not specifically discussed, however, would lose most of them.*
>
> *A mid-semester quiz, in DC machinery ... involved all dimensions of a motor being doubled. Several students said the test was "unfair" because we had not studied that. I personally learn most effectively by starting with simple examples and working up. Washington University was excellent in that professors such as yourself, Professor Glasgow, Dr. Bubb, or Dr. Middlemiss could and did take a student all the way back to the beginning if necessary and work up to the question at hand. ... I find that working with my children is a good testing ground for teaching methods because the subject matter is simple, the opportunities frequent, and the reaction clear.*

Frank Bubb was known for his innovative thinking and ability to challenge conventional methodologies. He encouraged students to look beyond textbook solutions and explore new ways of approaching engineering problems.

Ross Middlemiss taught a rigorous engineering mathematics course from 1929 to 1969. Walter worked hard to earn As in his math classes.

The relationships that Walter forged with his professors went beyond the classroom. Their influence provided him with a foundation that he would later use to reshape the field of control engineering. Although not a straight-A student, as Arline may have been, he earned A's in the majority of his classes, even English and history. Somewhat surprisingly, his lowest grades, "gentleman's C's," were in two mechanical engineering classes.[8]

John R. Moore: Mentor

At Washington University, Walter met John R. "Johnny" Moore, an intellectual peer who would become one of the most influential figures in his early career. These images from 1937 and 1941 were taken when they were seniors at Washington University. Johnny would become an important collaborator and professional ally. Walter would follow him to Schenectady in 1941, St. Louis in 1946, and Southern California in 1948.

That saga unfolds in later chapters.

John R. Moore in 1937, his senior photo at Washington University in St. Louis

Walter Evans in 1941, his senior photo at Washington University in St. Louis

The 1930s and 1940s were a period of profound growth in the fields of engineering and mathematics. Universities like Washington University played a pivotal role in fostering the next generation of innovators. Walter's time at Washington University coincided with a surge in research on servomechanisms, feedback control, and stability theory—topics that would later define his contributions.

Summary: The Roots of Root Locus

While root locus was still years away from being formalized, the foundational elements were all present in Walter's early life:

- **Intellectual discipline**, inherited from a family that prized education
- **A network of mentors**, who nurtured his engineering abilities
- **A partnership with Arline**, whose support provided stability

INTO STABILITY

- **The rigor of Washington University,** which taught him to think
- **John R. Moore,** who would later influence his professional trajectory

Each of these elements was a root in its own right. Together, they nourished the intellectual ground from which root locus would eventually emerge.

By the time Walter Evans graduated from Washington University in 1941, he had received more than an engineering diploma. He was also a thinker shaped by a rich lineage of scholars, strategists, and innovators who possessed the skills, relationships, and intellectual curiosity necessary to make groundbreaking contributions to control systems engineering.

Graduate Walter on campus of Washington University c. 1941

Walter's path was shaped by clear and distinct influences. His family, his mentors, his education, and his lifelong partnership with Arline were all essential elements of the equation. These were all roots of root locus—deep, interconnected, and essential to the innovation that would follow. As he embarked on the next stage of his journey, Walter carried these influences with him. His story would soon move beyond St. Louis, but its foundation would always remain rooted in the experiences, relationships, and institutions that had shaped his early years.

Chapter 2

Schenectady (1941–1946)

Upon graduation from college, Walter accepted a job offer from General Electric. In July 1941, he took up residence at Lake Lodge on Ballston Lake, New York, a few miles north of Schenectady. It appears his job did not begin until October, when he moved to an apartment on Bedford Road in Schenectady. His wedding to Arline Pillisch was set for April 11, 1942. Before returning to St. Louis, the twenty-two-year-old was apparently homesick, based on the letter he wrote to his mother.

Walter Evans and two colleagues at YMCA near Schenectady, NY

INTO STABILITY

Dear Mother,

Being away from home for a year is highly touted as a great experience for a boy in teaching him to cope with the responsibilities of life. Admittedly, I have learned quite a few facts, particularly concerning human behavior, but I really don't feel changed noticeably from last year in St. Louis. And most of the things I have noticed here in Schenectady recall conclusions that you had expressed before. The fundamental one is the same that struck Sammy too: most people are too embroiled in their own welfare to give much consideration to the general rightness of things. You easily make acquaintances among strangers, but genuine friendship requires time to develop. Please don't think from this that the going has been rough at all, but just getting back among St. Louis's friends is going to be a great experience.

Love, Walter

Arline Pillisch and Walter Evans, undated photo

SCHENECTADY (1941-1946)

On April 8, 1942, Walter carried his bags down through the boarding area of the New York Central and Hudson River Railroad station, his thoughts undoubtedly turned homeward. That he missed family and the "old gang" was evident in the opening the letter he had written a week earlier. His homecoming itinerary was tightly packed. Arriving at Union Station on Thursday, he planned to be "leading the pack down the stretch," as he put it in his letter. That afternoon, he hoped to relive a familiar ritual: biking the two miles from his childhood home on Nashville Avenue to Washington University, where he would visit two beloved engineering professors, Roy Glasgow and Frank Bubb. That evening, he and his mother were invited to dinner at the Pillisch home; Walter had asked Arline to tell her mother not to "outdo herself" in preparing the meal. Perhaps after dinner, there would be time to watch home movies with friends from St. Louis.

Friday would be a whirlwind of wedding preparations: securing the marriage license, renting his tuxedo, picking up Arline's wedding dress, purchasing train tickets, and meeting with the minister before the rehearsal dinner.

If the weather cooperated, he envisioned spending part of Saturday—his wedding day—taking a quiet drive with his mother in the family's "jolly old Oldsmobile," revisiting the outskirts of St. Louis. He acknowledged the wear on their tires, given that domestic tire production had been redirected for military use, but the nostalgia of seeing familiar sights was worth a few "precious miles." One potential stop was the airport, which was tied to Walter's courtship with Arline. In 1940, she had flown alongside him in a Piper Cub as he piloted them over their city—where they were born, where their parents were born, and where their story began.

High School Sweethearts

Their story stretched back to tenth grade geometry class at Soldan High School. Walter had loved geometry, and Arline had been drawn to his intelligence and wit. Their first and only high school date had been during their senior year, when Walter took her to see *The Taming of the Shrew*. Although she had dated others in college, Arline always had her eye on Walter. Every year, she waited for their one annual date: March 17, Engineer's Day at Washington University. As long as Walter took her out that day, Arline knew she was still his girl. Walter, ever practical, wasn't one for excessive sentimentality. However, he made at least one grand romantic gesture. The only surviving letter from him to Arline before their wedding was a belated Valentine's note from 1941, their senior year at Washington University. In it, he penned:

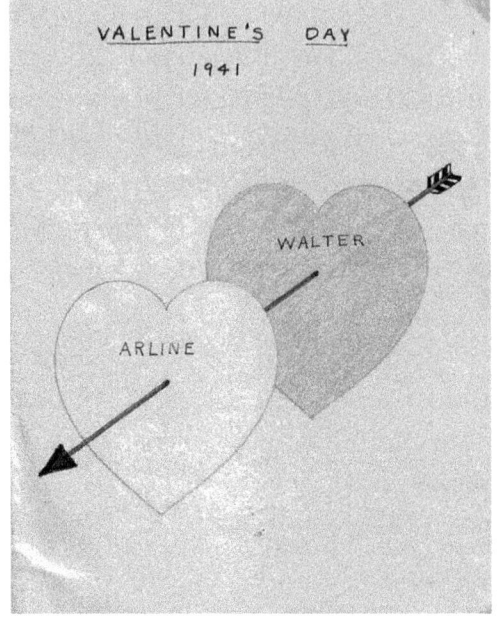

Will you be my Valentine?
Though the request be late.
I hope this oversight may prove
A lucky stroke of fate.

For if you like to keep a file
Of little things sweet and dear,
Please let this card have its place
To make just one point clear.

That men are prone to forget a date,
A time, a place, or occasion.
That a woman's heart is hurt thereby
Is a needless situation.

For if she knows, as know you must,
The strength of his affection;
Then in his love you may safely trust
If not in his recollection.

Walter P. S. I'll try to do better.

The note was classic Walter—self-aware, pragmatic, and dryly humorous. But beneath his understated style, the message was clear: His heart belonged to Arline.

Their high school principal, Herbert Stellwagen, would be an honored guest at their wedding. On the same Saturday afternoon that Walter had suggested he and his mom could spend time in the jolly Olds, Stellwagen would compose two letters of well wishes to the couple.

SCHENECTADY (1941–1946)

Dear 'Grandson' Walter: I am rejoicing with you and Arline today—on your happy wedding day,—in the joy of your lives so far, in the beauty of this day, and in all the glory to be in your lives together. How very much you both bring to your marriage; and to your helpfulness to each other; and to the new family which you will build; and, individually and together, to your achievement for yourselves; and to your service so greatly needed from you to society! ...

Dear Arline: While I have already written today to Walter for both him and you, I cannot refrain from sending this special note to you, to tell you how happy I am in the happiness of you and Walter today; how filled with admiration I am because of all you both bring of richest promise to your marriage that your lives together will be completely right because of the worthiness of the character, the health, the ideals, the ability, the attainment, and the will to serve that reside in each of you; and how sincere are my congratulations and best wishes to you both. ...

The bride on April 11, 1942, in St. Louis, Missouri

Schenectady

Walter and Arline's lives together would take them to their apartment at 117 State Street on the fiftieth anniversary of the April 15, 1892, merger of Edison General Electric and Thomson-Houston of Lynn, Massachusetts. This merger created General Electric Corporation.

During fair weather or foul, Evans would bike the two miles down Union Avenue and past Union College to GE's plant. The 600-acre grounds had 200 buildings. GE's Schenectady Works was ramping up from 29,000 prewar employees to its peak wartime workforce of 47,000. For most of the war, Walter's office was on the third floor of Building 22, near the entrance at the intersection of Weaver and Edison.

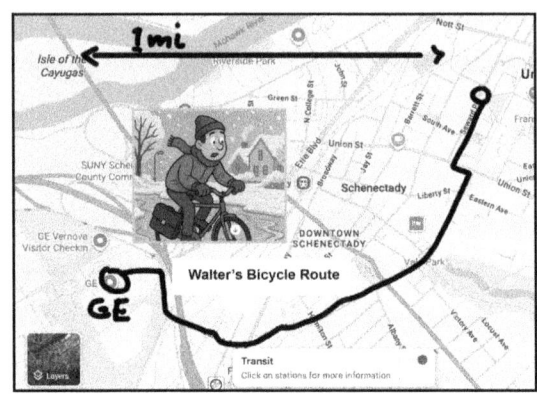

Evans's commute route—on his bicycle— fair weather or foul

Edison was named for General Electric's legendary founder Thomas Edison. He had founded the corporation fifty years earlier. Now Schenectady Works was the largest GE plant in the country. One gets a feel for the enormity of its wartime operations in this excerpt from *The Empire State at War*.

> *During World War II, the General Electric Company, whose general offices and largest manufacturing plant are located in Schenectady, was called upon to produce a greater variety of complex war equipment and to solve a greater diversity of difficult technical problems than any other manufacturing concern in the country. ... Twelve years' production was crowded into four. In the aggregate, the company turned out 4 billion dollars of war equipment, ranging from giant turbines for battleships to delicate instruments for airplanes and mass spectrometers for the atomic bomb project.*[9]

A large part of the complex war equipment produced by General Electric was made at Schenectady Works, the largest electrical workshop in the world. Many of the difficult technical problems of the war were handled here by the company's research and engineering laboratories. The expansion of these unique activities distinguished Walter's five years at General Electric from 1941 to 1946, combining practicing his craft, learning, and teaching.

Gordon and Rusty Walter: Friends and Officemate

Walter and Arline developed lifelong friendships during their time in Schenectady. One of Walter's closest bonds was with Gordon Walter, a valedictorian from Iowa State, who remained a steadfast friend for decades. The two engineers shared an office while balancing supervisory roles and completing their own coursework.

Walter and Gordon continued exchanging Christmas cards and letters. Gordon lived an extraordinary life, reaching 103, compiling more than 10,000 travel slides, and remaining active in his local church and community for decades.[10]

Walter fondly recalled the demanding schedules and harsh winters of Schenectady during World War II. Engineers worked six days a week, often pulling twelve-hour shifts, leaving little time for family or even homework. Bicycles were essential in the overcrowded city, as public transportation was unreliable.

Walter and Gordon were among those who braved the elements year-round, even when temperatures dropped below −20°F. The winter commutes were grueling, requiring multiple layers of clothing, thick mittens, scarves, and heavy headwear. Their legs, less protected, often took extra time to thaw once they reached the office.

Rusty and Gordon Walter
in Schenectady, NY c. 1942

An Unwelcome Letter from Home (October 1942)

On Saturday, October 3, 1942, Walter may have hoped for a rare moment of leisure, perhaps tuning in to the third game of the World Series. His beloved

St. Louis Cardinals had just secured a 2–0 shutout against the Yankees, putting them ahead in the series. But celebration was the last thing on his mind.

The previous day, he had received a heartbreaking letter from his mother: The family's beloved dog, Tigger, had passed away.

Tigger, the Evans family dog

That evening, Walter wrote back to his mother, struggling to process his grief. He reflected on the unique bond between humans and their dogs—how Tigger had always been full of life, watching over the household. Walter acknowledged that people often feel silly mourning a pet so deeply, but that didn't lessen the pain.

Hoping for a distraction, he and Arline went to see *The Pride of the Yankees*, expecting a lighthearted baseball film. Instead, they found themselves watching Lou Gehrig's tragic story unfold. The film's emotional weight broke through Walter's usual composure, and he found himself tearing up in the theater. "One good cry lowers your defenses for the next," he later wrote.

Back home, he found comfort in old photographs of Tigger—one of the dog tied in a canoe on the Merrimack River, another of him standing proudly in a creek at the family farm. He enclosed one of his favorite pictures in the letter, imagining Tigger at peace: "If all life's stories were as simple and fulfilling as Tigger's, the world would be a better place. We gave him a happy home and, in return, he gave us the pure joy of his unshaken loyalty."

John and Miriam Moore: Friends

Beyond their friendship with Gordon and Rusty Walter, Walter and Arline also grew close to another couple during their GE years: John and Miriam Moore. This bond extended beyond mere social connection; Walter followed Moore's career trajectory at multiple key points. He first joined him at General Electric, later returned with him to Washington University, and eventually followed him again to North American Aviation. Both men made significant contributions to servomechanism analysis; both published influential research and balanced teaching with hands-on engineering.

John and Miriam Moore with Arline and Walter Evans near Schenectady, NY, in 1945

In a 2003 interview, Moore spoke warmly about his time with Walter, recalling the fun they had composing new lyrics to "Love Life of an Engineer," a song set to the tune of "The Battle Hymn of the Republic."[14] He then, unprompted, regaled the author with a spirited, spontaneous rendition of all twelve verses of the song, which had originally been written at the University of South Dakota.[11] The ballad humorously captured the struggles of engineers. Moore credited Walter with crafting particularly witty verses:

> *"His quick response to error was the follow-up to this tale,*
> *But marriage proved too steady state to match his transient flare.*
> *So boys take heed, let this sad yarn be a warning that runs deep—*
> *Don't build a power plant where local power's cheap."*

Chapter 3

The GE Advanced Course (1941–1945)

The primary attraction for Walter when seeking employment at General Electric Corporation was its famed Advanced Course, an elite training program for promising engineers. Washington University had developed a tradition of its top engineering graduates applying for entry into the

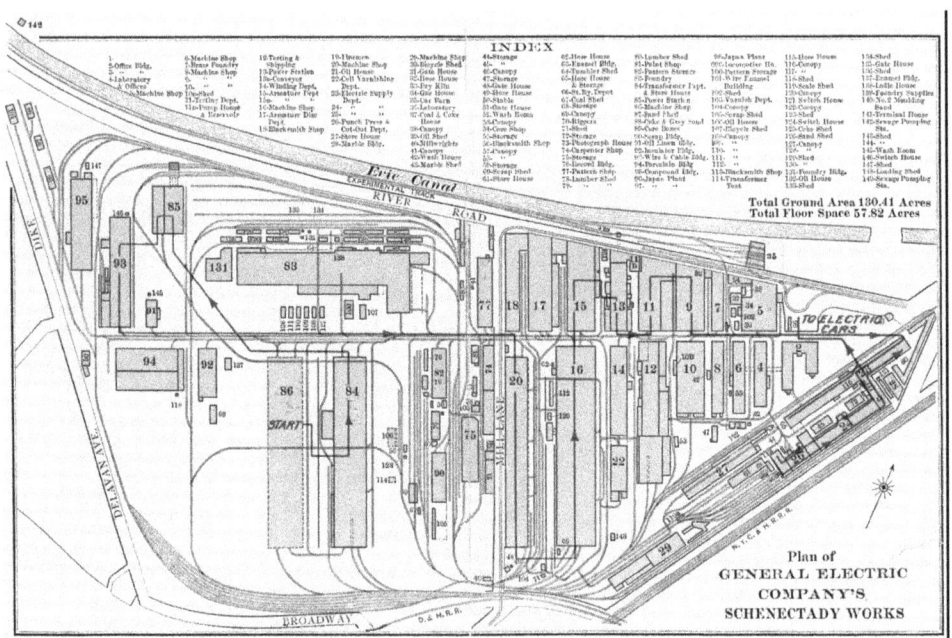

General Electric Schenectady Works Plant Map

program, and Walter followed in the footsteps of several distinguished predecessors. In 1941, he was one of two Washington University graduates to be accepted into the program's A Class, alongside N. A. Schuster. That year's cohort included single representatives from Ivy League institutions such as Yale, Princeton, and Brown, as well as renowned technology institutes like MIT, Caltech, Rensselaer, and Carnegie Tech.

The Philosophy of the Advanced Course

The educational approach of GE's Advanced Course would have a profound impact on Walter's views of engineering and learning. In 1941, the program's grand overseer was Alexander Stevenson Jr., an accomplished engineer whose earlier work had led GE to manufacture electric refrigerators. Stevenson articulated the purpose of the Advanced Course in a 1935 paper, describing a need for engineers with deeper analytical skills:

> *Most [engineering graduates] had no mathematical training beyond calculus, and those who had studied differential equations had not learned how to make use of them in the analysis of physical problems. There were coming to the company many excellent, ingenious engineers, but ... courses in many colleges were being taught routine rule-of-thumb methods of design, and there was insufficient emphasis on thinking problems through by the use of fundamental principles.*[15]

The true architect of the Advanced Course, however, was Robert E. Doherty, a nationally respected educator who had designed the program in 1922. By 1941, Doherty was the president of Carnegie Institute of Technology, and his vision for engineering education was deeply embedded in the course's philosophy. Walter kept in his Advanced Course notebook a copy of Doherty's address to students, which emphasized the importance of true understanding over rote memorization:

> *So, I urge you to take the initiative and learn to use your heads. In the first place, dig yourself out of confusion. Insist on understanding! Do away with superficiality! Stop memorizing words and formulas that you don't understand, merely for a grade.*
>
> *Don't go on cultivating a habit that will cripple your mind for the rest of your days—the habit of superficiality, the habit of playing on words*

that carry no meaning. You know when you understand and when you don't; when you grasp a point that is clear and clean-cut and when, instead, it is blurred and confused.

With all the emphasis in me, I repeat: insist on understanding! Then, under the guidance of the faculty in your regular class programs, but under your own initiative, you will be in a position to go forward more effectively with the acquaintance of a genuine education gauged to the demands of the changing world in which you will live.[12]

Doherty's emphasis on deep comprehension rather than superficial learning was rooted in his own formative experiences at General Electric. By 1910, he had come to appreciate the power of conceptual clarity thanks to his mentor, Charles Steinmetz—arguably the most brilliant mind ever to work at GE. At a May 1942 banquet for the Advanced Course, which Walter likely attended, Doherty reflected on his early interactions with Steinmetz.

He recalled how, as a young engineer, he often sought Steinmetz's help when encountering difficult problems. Instead of providing direct solutions, Steinmetz would reference a fundamental physics principle that held the key to the answer. Doherty already knew these principles—he could recite them from memory—but he had not yet learned to *apply* them. His struggle was rooted not in a lack of knowledge, but rather in a lack of analytical thinking. He had been searching for formulas where none existed; what he truly needed was a conceptual framework for problem-solving.

Doherty also spoke about the broader vision behind GE's Advanced Course: assembling the brightest engineering minds not just to learn, but also to form lasting professional networks. He saw immense value in creating an informal community of engineers who could rely on one another. If one of them got stuck, they wouldn't be alone. They would know exactly whom to turn to for guidance.[17]

Friends at GE in an Advanced Course Study Session

A-Class Year (1941–1942)

In September 1941, Walter and Gordon were selected for one of two parallel A-Class programs, a rigorous training initiative blending classroom instruction with hands-on experience. The program followed an eight-point framework established in 1935 by Alexander Stevenson Jr., who believed early-career engineers needed to develop both theoretical knowledge and problem-solving skills. The A-Class emphasized applying core engineering principles—mechanics, thermodynamics, heat transfer, electricity, and magnetism—to real-world challenges. Students weren't just solving textbook equations; they had to sift through raw data, make educated assumptions, and think critically. The overarching goal was to instill a problem-solving mindset that transcended formulas, using the following framework:

Evans studying at General Electric, circa 1941

1. Recognize that fundamental physical laws apply across all disciplines.
2. Learn to think independently and use these laws to solve problems.
3. Identify relevant data from a sea of available information.
4. Strengthen mathematical reasoning in practical applications.
5. Simplify complex problems using sound engineering judgment.
6. Persevere to work through difficult problems step-by-step.
7. Cultivate self-criticism and the ability to validate one's own answers.
8. Communicate ideas clearly, concisely, and persuasively.[15]

A-Class students worked full time in test divisions, but not every assignment was stimulating. In late 1941, Walter found himself stuck in a monotonous calculation test job in Building 258. Frustrated, he wrote a five-page critique of the assignment, arguing that the country and GE needed engineers prepared for greater responsibility:

A dissatisfied test engineer—or one who merely endures his three-month assignment—is hardly a strong candidate for any engineering department. At this critical moment, both the country and General Electric need engineers ready to step into responsible positions. Assigning trained engineers to work that could be done just as well by someone without their background is a needless waste.

His frustration was justified; the workload for A-Class students was immense. In a 2003 letter to the author (see Appendix 3), Gordon recalled,

Class members were holding full time jobs during their time of participation. The only time away from those jobs during working hours was for the four-hour class period once a week. The problems were solved at home and, for the A Class, could require between 20 and 25 hours a week. Most of this time was needed to get some understanding of the basic principles involved and to express them correctly mathematically. The rest of the time went into trying to solve the resulting mathematical expressions.

For Walter, the A-Class experience laid the foundation for his career. The problem-solving approach he honed during those years influenced his groundbreaking work on servomechanisms, leading to his 1948 and 1950 AIEE papers and his 1954 book, *Control-System Dynamics*. The lessons learned in Schenectady shaped his entire approach to innovation and analysis, which would leave a lasting impact on the field of engineering.

B-Class Year (1942–1943)

Aside from Walter's B-Class photograph, little documentation remains from his second year in the Advanced Course. According to Stevenson's original framework, those advancing to the B-Class left the testing department and became full-time employees of the program. This shift allowed them to take on real engineering assignments lasting three to four months. They gained hands-on experience in different departments while continuing their classroom studies for half a day each week.

Walter received a modest pay raise—from seventy-five to eighty-five cents an hour—and transitioned into the Advanced Course program office, where he rotated through assignments from 1942 until the spring of 1943. His supervisor was Harold Chestnut, an MIT graduate just two years his senior. By 1940, Chestnut had earned both his bachelor's and master's degrees in electrical engineering; he was deeply involved in GE's Aeronautic and Ordnance Systems division. This division worked closely with MIT's Servo-Mechanism Laboratory, led by Gordon S. Brown, and the Radiation Laboratory, directed by N. B. Nichols, on the gun turret control systems for the B-29 bomber. Given Chestnut's role, Walter was probably introduced to servomechanism analysis—a field that would define his career—during this period.

A-Class Supervisor (1943–March 1945)

Alongside their enrollment in the C-Class, both Walter and Gordon were appointed as full-time supervisors in the Advanced Course. Walter was assigned to oversee an A-Class, while Gordon supervised a B-Class. They moved into a shared office, marking the start of a professional partnership that would last for years.

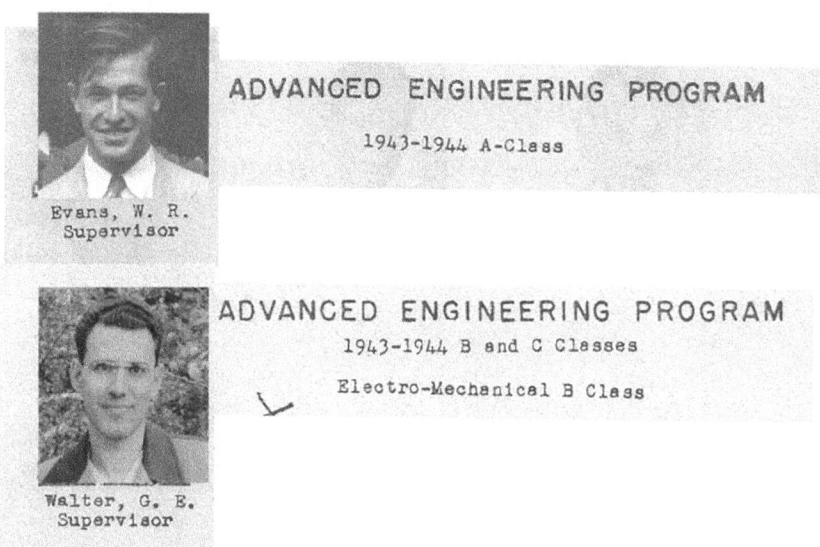

Walter took his supervisory role seriously, meticulously documenting his students' progress. His personal binders contained everything from introductory remarks for guest lecturers to detailed performance charts. On the first day of class, he gave his students a straightforward message:

> *My role as supervisor is to help each of you get the maximum possible benefit from this program. I'm fortunate to build on the experiences of previous classes, but the course itself is always evolving. We make*

changes every year in an effort to improve. One conviction I hold strongly is that honesty is key—so I encourage open discussion about every aspect of the program.

Beyond exams and coursework, students were required to propose a project idea. Some suggestions were remarkably forward-thinking:

- J. L. Knox: Television recorder and playback
- H. Gayek: Electronic metronome, thermostatic shower control
- J. A. Massingill: Run-less silk stockings, unbreakable thermos bottle
- P. A. Thompson: Vacuum plumbing system for home cleaning
- J. Townsend: Automatic pinsetter for bowling alleys
- J. J. Zaskalicky: Atomic power (in 1944!)

A Disrupted Program

As Walter and Gordon settled into their new supervisory roles, the Selective Service System began reassessing whether to continue granting occupational deferments to engineers working in Schenectady. The war effort was intensifying. Draft boards questioned whether more engineers should be called into military service, despite their work contributing to national defense.

Then, on March 25, 1944, General Electric made a stunning decision: All Advanced Course classes were suspended indefinitely. Students scattered—some returned home, others enlisted in the military, and a few remained at GE in various engineering roles. The program wouldn't resume for more than a year. In the meantime, Walter and Gordon attempted to maintain a sense of continuity through correspondence, hoping to keep the spirit of the program alive.

Brotherly Encouragement: You Are Doing Your Part

On June 6, 1944, the United States and Great Britain, under the leadership of General Dwight D. Eisenhower, launched the greatest naval armada in history, crossing the English Channel to liberate Europe from Nazi control. Walter's brother Sam, a captain in an engineering command, was responsible for laying gasoline pipelines to supply American troops advancing through France toward Germany.

Shortly before D-Day, Sam abruptly canceled a trip he and his brother had planned. Due to military security, he never explained why. Afterward, Walter wrote Sam a letter. In response, on June 16, Sam wrote back, apologizing for the sudden cancellation. He also expressed his joy at the birth of Sam's first son, Richard (Dick) Arthur, on May 31, writing, "It is really nice to just know that you have a son."

Sam then addressed an issue Walter had likely raised in his letter: the relative value of their contributions to the war effort, given that both had received Selective Service deferments for their technical work:

Capt. Sam Evans, circa 1944

> *I received while on the boat the letter that you wrote on that eventful Sunday. Gee, but it was good to read, too. I went over it a couple of times. … You shouldn't feel any lack of pride in your work. It's essential and contributing to the war effort, so you are doing your part. … When this whole thing is over, I don't plan to discuss it a great deal with my children … I do believe the subject should be discussed enough to develop an international organization to delay a recurrence of a world war in the immediate future.*

Fifty years later, Sam returned to France with his children and grandchildren to retrace the steps of his company from the Normandy beaches. He estimated that he and his men had laid 350 miles of pipeline. In the French villages they visited, elders welcomed them with smiles and gestures. One family proudly showed them a piece of the pipeline Sam and his men had installed, which they had incorporated into their home as a keepsake. Reflecting on her father's generation, Sam's daughter Peggy remarked, "It's amazing what these men did."

Aeronautic and Ordnance Division (1944–1945)

In December 1944, during a pause in his training duties, Walter conducted a self-assessment of the A-Class program. He listed all lecturers, most of whom were recent graduates of the Advanced Course. Walter himself gave eight lectures, Gordon gave two, Alexander Stevenson gave one, and Simon Ramo spoke once. While generally satisfied with the program, Evans identified areas for improvement, particularly in maintaining its original focus:

> *The A-Class achieved distinction by its concentration on a few fundamental principles, thoroughly understood. As the program of topics develops, higher mathematics and technical information are added. Each supervisor, perhaps influenced by subconscious pride, is tempted to push a topic just a little further than before. … It might be well for each supervisor to consciously question each point of his proposed program: Is this topic concerned with the application of fundamental principles to engineering problems?*

In the spring of 1945, Walter may have been assigned to work under Harley Bixler. If so, this meant that servomechanism design and analysis would have become Walter's full-time focus. One of his key projects involved designing a remote-controlled gun turret—a classic servomechanism problem. The system compared the direction in which gunners aimed their sight with the actual angle of the gun; it used motors to drive the error signal to zero.

A-Class Resumption (July 1945)

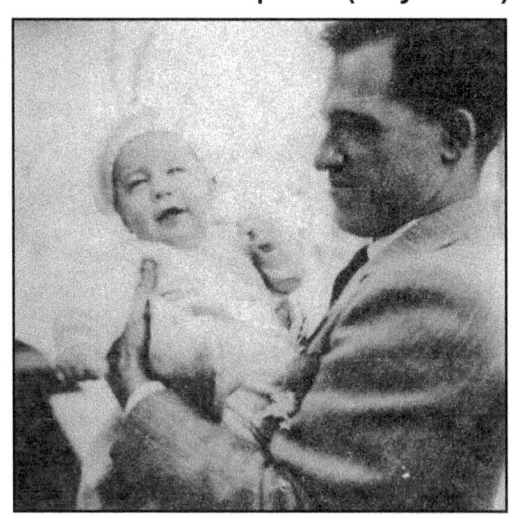

During the yearlong suspension of all classes, Walter kept his A- and B-class students apprised of events, including a personal one—his and Arline's first child in November 1944, Randall Gomer Evans. (Gomer was Walter's father's, grandfather's, and uncle's first name.) "A son is now rapidly demanding

Randall Gomer Evans and proud father, 1945, Schenectady, NY

more attention in the Evans family. At eight months, he seems more destined to become a fighter than an engineer, as his curiosity takes the form of beating on everything. I'll be glad when he has my nose figured out."

On May 5, 1945, anticipating the program's possible resumption of B-Classes in October, Walter sent a letter to Section I B-Class students inquiring about their experiences over the past year and whether they would be interested in returning to General Electric for Section II of the B-Class. He assigned them a homework problem humorously entitled, *II B or Not II B*. Remarkably, every student responded with a multipage handwritten letter.

To his A-Class students, Walter revealed that General Electric had decided to further delay their advanced engineering program. "The next news announcement will be made approximately November 1, unless something significant develops before then," he wrote. And something did: peace.

In April 1945, Nazi Germany surrendered unconditionally following Hitler's suicide, abruptly ending the war in Europe. Four months later, the United States dropped atomic bombs on Hiroshima and Nagasaki. Japan surrendered.

Post–World War II (1945–1946)

With the war's end in August of 1945, General Electric restarted the Advanced Course on October 1. Walter became increasingly engaged in its administration. He may have supervised a section of a C-Class, though scant evidence supports this. Personal records confirm that during this final year at General Electric, he devoted significant thought to career-related questions. Would he choose to follow his father and brothers into engineering administration or pursue a technical career path?

As an A-Class supervisor in 1945, Walter had encountered various engineer attitudes. He retained a thick compilation of "class criticisms" and developed a growing interest in improving company programs. This led him to join a committee of five Advanced Course engineers tasked with offering constructive criticism of the Test Program. He joined the Schenectady General Electric Engineers' Association (SGEEA) and helped develop a questionnaire for engineers, receiving more than a thousand responses. This survey included both attitude and salary components. These actions suggest that he considered a management career path.

Evidence that Walter would choose to follow a technical career path lies in his avid pursuit of more knowledge about servomechanisms and, in particular, the advances at General Electric. He became frustrated by the apparent lack of documentation. Consider this September 1945 complaint:

> *A trip to the Data Bureau to find Data Folders or Technical Reports on the subject (of servomechanisms) produced only one report, and that was just because it had been just issued. The difficulty seems to be that maintaining the wealth of information has been relegated to one man as a part-time job. Now that the war is over, can't more attention be paid to properly indexing the General Electric technical knowledge?*

SCHENECTADY ENGINEERING COUNCIL

AIEE · ASME · NYSSPE · AWS

Bulletin

ASCE · ACS · SAE · ASTE

VOL. III   NOVEMBER, 1945   No. 3

New Challenges For Engineers

by WALTER R. EVANS, *Associate Editor*

The challenge of war is definite: to win. The engineers may all be proud of their record of development of amazing military equipment. One may well ponder in retrospect what were the factors which contributed to this achievement. Certainly one of them is the mere fact that the purpose of our efforts was clear-cut. The profession is now free to accept new responsibilities. But the challenges of peace are not so definite. We need the stimulus of discussion to realize the many problems which await solution. The purpose of this article is to set forth in very general terms three problems, which new devices may help solve: maintaining full employment, stimulating long-range scientific thinking, and speeding the processes of democracy.

He followed up with an article, "New Challenges for Engineers," published in November in the Schenectady Engineering Council *Bulletin* (see Appendix 2).[13] Then, in 1946, Walter kept notes on topics such as "Matching Men and Their Jobs," "Classification of Engineers," and "How to Determine Employees' True Feelings Toward Company Policy." He attended meetings on engineering approaches to human problems and workforce retention. The company valued his input. In April 1946, they sent him on a week-long tour of four General Electric works—River Works, Thomson Laboratory, Bridgeport, and Fort Wayne—to explore potential career opportunities. He thanked his superiors for giving him "a free hunting license in seeking the best possible assignment."

Amid these activities, Walter prepared a paper for a General Electric engineering competition. Entitled "Let Horsepower Do the Horsework," it proposed new applications for IBM's wartime calculating machines, which were largely confined to accounting. He suggested using punch cards to store engineer ratings and comparing them against career trajectories of successful engineers.

In summary, evidence exists for Walter's intention to pursue a management career and technical career paths. However, toward the end of 1946, he discovered a paper entitled "A New Method for the Treatment of Regulation Problems" by a Swiss mathematician, Paul Profos, published in an obscure journal, *Sulzer Technical Review*. This discovery may have nudged him toward the technical. In the words of poet Robert Frost in "The Road Not Taken," Walter's discovery of the Profos paper may well have "made all the difference." The next chapter begins with Gordon Walter's account of events. In the words of the more modern poet, Lin-Manuel Miranda, Gordon was "in the room where it happened."

No. 2 **SULZER TECHNICAL REVIEW** 1945

A New Method for the Treatment of Regulation Problems*

Paul Profos introduced a combined graphical and computational technique to evaluate system stability and natural vibrations. He emphasized its value in applications such as temperature regulation in steam boiler superheaters and heating plants, where standard control theory was insufficient. Tests confirmed the method's practicality, pointing to wide potential for further use.

Paraphrase of original by ChatGPT

Chapter 4

The Graphical Analysis Paper (1946–1948)

Swiss Mathematician Professor Paul Profos (June 1946)

From 1943 to 1946, Gordon Walter shared an office with Walter Evans. In a 2004 letter, Gordon recounted a pivotal moment in Walter's intellectual journey:

Walt and I spent quite a few evenings together in the otherwise deserted Advanced Course office, preparing material for our students while managing our day-to-day jobs on B-20 fire control. One evening, Walt began reading a paper by a Swiss mathematician on a different way to analyze the roots of a polynomial equation. He was intrigued, spent more time thinking about it, and eventually developed an interest in applying it.

We had been exposed to feedback theory in Advanced Course classes, and Nyquist diagrams [Harry Nyquist] were in vogue. Our assignments involved servo systems. For whatever reason, this was when Walt caught the bug that led to root locus analysis and the Spirule. He had the insight, determination, and ability to take the initial concept to a higher theoretical and practical level.

I just happened to be in the same room when the bug first bit him.

The Swiss mathematician was Professor Paul Profos, whose 1945 article "A New Method for the Treatment of Regulation Problems," published in *Sulzer Technical Review*, arrived at the Schenectady library in June 1946. Walter was captivated by it, as it added a new dimension to Nyquist's frequency response method.

Going Home (September 1946)

Shortly thereafter, Walter and Arline made a career-defining decision: Walter would leave General Electric for a faculty position at Washington University in St. Louis. John and Miriam Moore had already decided to leave Schenectady, citing their daughter's health issues caused by the harsh winters. The winter of 1945–1946 set a record for consecutive freezing days (thirty-six) that still stands.

Many factors likely influenced the Evanses' decision, with family and familiarity playing a significant role. "You easily make acquaintances among strangers, but genuine friendship requires time to develop," Walter had written to his mother in 1942. That sentiment surely remained true.

But the man who returned in 1946 was a changed one. Walter was now a husband and father. The world had witnessed the defeat of the Axis powers, the death of Franklin Roosevelt, and the dawn of the atomic age. Professionally, he had mastered servomechanisms, taught classes, and discovered Profos's novel analysis. As an instructor at Washington University, Walter would have the opportunity to further his studies, engage with students and faculty, and, perhaps, publish a paper inspired by Paul Profos's graphical techniques.

In September 1946, Walter, Arline, and their son Randy boarded a train at Schenectady's Union Station and returned home to St. Louis, closing one chapter and opening another.

5234 Lotus Avenue, St. Louis, Missouri

Upon returning home to St. Louis from Schenectady, the Evans family moved into the second floor of the house where Arline had grown up—5234 Lotus Avenue—sharing it with her parents, Olinda (née Meyer) and Reinhold Pillisch. Also living there was Arline's sister, Eleanor, who was unmarried at thirty-two. That status would change a few months later.

Arline's father was a postal carrier (then commonly referred to as a "mailman"), while her mother was a homemaker. Neither of them ever held a driver's license, so they relied on public transportation to get around the city, as did Arline.

Their home on Lotus Avenue, located north of Forest Park, was just a manageable three-mile bicycle ride from Washington University's hilltop campus on the west side of the park.

Records indicate that Walter taught one graduate course on servomechanisms, using the gun turret system he had worked on in Schenectady as examples. He reported to his former professor, Roy Glasgow, a man known for his humor. Walter later recalled one of Glasgow's clever aphorisms, "The vector sum of all opinions is zero."

By 1945, Walter had learned the Nyquist and Bode servo diagrams at GE when he had been assigned to a group tasked with the design of a remote-controlled gun turret. He had also studied Dr. Profos's article "A New Method of Regulating System Design." Profos's method involved a curvilinear grid based on a Nyquist curve; he used the grid to determine an eigenvalue at the minus-one point. Walter found this idea compelling and continued exploring it while in St. Louis from 1946 to 1948.

Walter's friend and colleague John R. Moore had also returned to Washington University and was busy setting up a Mechanics

5234 Lotus Avenue, St. Louis, Missouri

Brookings Hall, Washington University in St. Louis

Laboratory. Walter's mother still lived in Walter's Nashville Avenue boyhood home. His older brother Cedric remained in St. Louis, working as a manager at Wagner Electric. His brother Sam, who had since entered the oil business, had moved away. His sister Alice, who had married an army officer, Duncan Hallock, had relocated as well.

Living with Arline's parents on Lotus Avenue provided the young couple with an opportunity to save money for a future home. With their second child on the way, they also had free childcare for Randy.

Walter and Arline likely reconnected with old friends from their high school and college years, though specific details aren't documented. What is certain, however, is that Walter's fascination with Profos's ideas about conformal mapping in the complex plane deepened. At Washington University, he would have found colleagues willing to discuss and critique his and Profos's ideas.

When Walter believed he had developed a sufficiently novel graphical analysis method to merit publication, he wrote to C. F. Wagner, secretary of the Committee on Servomechanisms of the AIEE, and inquired about submitting a paper.

Letter to C. F. Wagner (December 9, 1946)

The purpose of this letter is to learn the procedure for submitting a paper. The proposed title is "Graphical Analysis of Servomechanism Equations." The idea, believed to be new, is the extension of the Nyquist or Bode diagrams to include the damping constant as well as a real frequency. The usefulness of the method is that the transient response of a system is directly determined rather than merely implied by the steady-state performance. ... The next meeting listed is not scheduled until June, so I would appreciate knowing if there is an earlier meeting where this paper might be appropriate.

Walter continued teaching while refining his paper. In June, he took a summer job at Emerson Electric, arranged for him by Roy Glasgow. During lunch breaks, he befriended Ira Lohman, a non-engineer with whom he would maintain a lifelong friendship. (Years later, Walter's grandchildren, Thomas and Stephen Evans, would visit the Lohmans at The Forum at Rancho San Antonio.)

INTO STABILITY

In July 1947, Walter submitted his paper, hoping for publication in November.

Letter to Charles S. Rich (July 30, 1947)

I wish to submit a paper for presentation at the AIEE Midwest Convention in Chicago, Illinois, in November. The title is "Graphical Solution of Servomechanism Equations." The abstract is enclosed. The paper will be close to the six-page limit, as much space is needed for the dozen figures and numerous equations, which experience has shown to be effective in explaining the method.

Walter also sent a letter to Paul Profos, expressing appreciation for his work.

Letter to Paul Profos (August 11, 1947)

Your idea of using the Nyquist plot as a single line of a conformal map to determine the complex root of the differential equation seemed to me to be the foundation for a very practical method.

In September, 1946, I became an instructor of Electrical Engineering at Washington University in St. Louis, Missouri, and worked off and on with your idea during the fall and winter ... I believe you made a real contribution to the analysis of regulators, and I hope this paper succeeds in calling attention to it.

On August 16, Arline gave birth to their second child, Gregory Walter Evans.

Walter resumed teaching in September. In October, he received his first rejection letter: a moment that foreshadowed future battles with reviewers and publishers.

Evans family in 1947:
Arline, Walter, Gregory, and Randy

THE GRAPHICAL ANALYSIS PAPER (1946-1948)

Letter from Charles S. Rich (October 17, 1947)

We have been badly congested. Your paper was carefully reviewed, but one reviewer, well-versed in servomechanisms, reported difficulty in understanding your approach. He felt it contained unnecessary material and lacked focus. While "the subject has potential, it would require rewriting rather than revision."

Undeterred, Walter responded diplomatically, seeking clarification.

Letter to Charles S. Rich (October 21, 1947)

Your letter of October 17 puts me in a difficult position. The few specific criticisms suggest a misunderstanding of the main idea. Staff members new to the subject grasped the paper, but those trained in existing techniques struggled to break free from their framework. I would appreciate knowing if the Basic Science Committee was involved in the review process. If so, I can be assured of diverse perspectives in the next round of review.

Letter from Charles S. Rich (October 24, 1947)

I cannot add or detract from the original comments of the reviewers. The comments were a bit terse, and one of the reviewers is known to be a severe critic.

The Committee on Servomechanisms, which reviewed the paper, is a joint sub-committee comprising members of the Committee on Basic Sciences, the Committee on Instruments and Measurements, and Committee on Industrial Controls. We hope you will see your way clear to revise the paper, as the reviewers indicated it could be developed into a good paper through rewriting.

Walter worked diligently over the following weeks to revise the paper. He may have set an internal deadline of November 14 because the very next day, Arline's sister Eleanor married Elmer Biskup, a leather quality control worker for the Brown Shoe Company. Elmer was six years younger than Eleanor. Walter and Arline served as best man and matron of honor for the ceremony on November 15.

> GRAPHICAL ANALYSIS OF CONTROL SYSTEMS
>
> WALTER R. EVANS, ASSOCIATE AIEE
>
> ## Synopsis of Graphical Analysis Paper (1948)
>
> SYNOPSIS
>
> Walter R. Evans introduced graphical techniques for predicting transient responses in control systems by locating the roots of the characteristic equation. He connected this root-based approach to Nyquist plots, showing how damping and oscillation could be visualized on the complex plane. The method offered engineers a practical way to estimate system behavior from empirical data, openng broad possibilities for further development.
>
> Paraphrase of original by ChatGPT

Letter to Charles S. Rich (November 14, 1947)

> *The revised version of the paper, "Graphical Solution of Servomechanism Equations," is enclosed under the new title: "Graphical Analysis of Control Systems." The change was made to eliminate the implication that only servomechanisms with known equations can be solved. The method applies to any control system, even those for which characteristics are only known through empirical data. I would like to present the paper at the Winter General Meeting in Pittsburgh, Pennsylvania.*

Acceptance! (December 29, 1947)

> From: Charles S. Rich (via Western Union) *Graphical Analysis of Control Systems Accepts for Winter Meeting and Transactions. Scheduled in Sessions on Servomechanisms.*

It was approved by the AIEE Technical Program Committee for presentation at the AIEE Winter General Meeting (January 26–30, 1948).

THE GRAPHICAL ANALYSIS PAPER (1946-1948)

Feedback Evans Received After the Paper Was Published
April 22, 1948 and April 27, 1948: From G. Ross Henninger, Editor

> ... Incidentally, it may interest you to know that your paper promises to be the most popular single PROCEEDINGS Section on the April Order Form.

> ... It was gratifying to learn that it promises to be the most popular of the April order form. The lack of written discussion had frankly confused and worried me since the basic idea of the paper seemed to me to be very significant.

A Momentous Development in California (Summer 1947)

Months before, while Walter was still preparing his paper for submission, events unfolding two thousand miles away would profoundly impact his career—perhaps even more than the recognition his "Graphical Analysis" paper would receive. John Moore later described events that would alter the course of Evans's life:

> In the summer of 1947, Dr. Simon Ramo, a former GE colleague, invited me to Los Angeles to discuss a position at Hughes. While in California, I ran into Lynn Gore, another former GE associate, now working at North American Aviation's (NAA) newly formed Aerophysics Laboratory. There, I met Dr. Bill Bollay, who led the lab. He had just hired Dr. N. E. Edlefsen as Director of Guidance and Control and mentioned my name to him. Soon after, Dr. Edlefsen offered me a position as head of his Electromechanical Group, overseeing about 40 engineers. I chose North American Aviation rather than "getting involved with Howard Hughes,'"whose reputation for instability was well known. [19]

Instability was also a well known and serious phenomenon for control systems. Moore's decision to accept Dr. Edlefsen's offer set off a chain reaction of events culminating a year later with control system designers enabled, for the first time, to achieve *stability by design*. That story unfolds in the next chapter.

Chapter 5

North American Aviation in 1948

The end of World War II had precipitated a bust to the aviation industry boom years of 24/7 factory aircraft assembly lines. Industry executives realized they would have to reinvent their companies to survive. Companies like North American Aviation (NAA), which had thrived on the wartime demand for bombers, fighters, and trainers, were suddenly faced with the challenge of peacetime adaptation. Southern California, a hub for aircraft manufacturing during the war, now teetered on the edge of economic uncertainty as thousands of skilled workers and engineers wondered what the future held.

For many aerospace firms, survival hinged on diversification and innovation. Some transitioned to civilian markets, producing commercial aircraft or leveraging their expertise in other industries. Others sought opportunities in the nascent fields of guided missiles, jet propulsion, and experimental technologies. North American Aviation distinguished itself during this period by taking bold steps into uncharted territories of aerospace engineering.

Dutch Kindelberger and Lee Atwood at NAA understood that this new industry, to become known as "aerospace," would be founded on technologies that World War II had either jump-started or promoted to a new level. These included radar, propulsion, materials science, inertial guidance, and automatic control systems based on servomechanisms.

Applying these wartime advances to their business would require retraining their workforce and college hires, most of whose textbooks bore prewar publication dates.

North American Aviation's Aerophysics Laboratory

With this lofty goal, NAA established its Aerophysics Laboratory to spearhead advancements in aerodynamics, propulsion systems, and materials science. This farsighted act set NAA apart from competitors. NAA not only maintained a foothold in the shifting defense industry, but it also positioned itself as a pioneer in the exploration of supersonic and even spaceflight capabilities. By investing in theoretical research and experimental projects, NAA attracted top scientific talent and forged a path toward groundbreaking achievements that would define the aerospace industry for decades.

At the beginning of 1948, Walter was in his second year of teaching a graduate-level class in servomechanisms at Washington University. His family continued to live with his in-laws, where his wife cared for their two children—Randy, age three, and Gregory, five months old.

In late January, Walter presented his now-famous paper. Only a week later, John Moore began his tenure at NAA and immediately set to work recruiting talent for the Aerophysics Lab. Walter must have been high on Moore's list. Moore described his vision for the team and the qualities he sought in its members:

> *… we set about building an organization which was to become unique among the electronics and control organizations of the defense industry. This was because we concentrated to the maximum extent on hiring generalists, extroverts and entrepreneurial types. The reason for this was that we were seeking leaders with drive and personal stamina, who could identify the important issues and recognize points of diminishing returns rather than solving the wrong problems very accurately.*
>
> *Much of this early hiring was done out of universities, where many of the grads were World War II veterans, attending on the GI Bill. We ultimately ended up with so many outstanding professionals that suffice it to say we had top stars among our industry in all aspects of our work.*[19]

Sometime after February, Walter accepted an offer for the summer of 1948 that would begin immediately after he fulfilled his teaching responsibilities. He planned to fly to California, live in an apartment, and work at NAA while his wife and children remained in St. Louis with her parents, sister, and brother-in-law.

Walter's primary summer assignment at NAA would be to teach employees the same graduate-level servomechanism class he had taught at Washington University. The course would begin in July. His Santa Monica apartment, located on the Pacific Coast, felt like heaven compared to the oppressive heat of St. Louis summers and the harsh winters of Schenectady. Even more inspiring was his audience: professional engineers who were deeply interested in the subject and eager to apply it to their work.

Another factor that may have played a role in his decision to leave Washington University was the arrival of a new university chancellor and dean of the School of Engineering. They would prohibit the engineering faculty from augmenting their salary during the summer doing consultation for outside companies, which would greatly reduce their income. Professors Glasgow and Bubb resigned and were now employed elsewhere.[*]

The city of St. Louis may have reached its zenith as the "Fourth City" in the United States during the early twentieth century, particularly with the hosting of the 1904 World's Fair and the Olympics. These events showcased St. Louis as a vibrant, growing hub of culture, commerce, and innovation. The rest of the twentieth century was less kind to St. Louis, as it was to many other cities in the Midwest.

Westward Ho!

These factors, combined with John Moore's persuasive power, led to Walter and Arline's joint decision for him to accept a permanent position with NAA rather than return to St. Louis at the end of the summer. The move marked a pivot in Walter's professional trajectory. NAA offered him both a platform for research and a technically sophisticated audience to test, critique, and

---

[*] Dean of Engineering Alexander Langsdorf retired in 1948. Dr. Arthur Compton, recently hired as the new university chancellor, appointed Lawrence Stout the new dean. Stout served until Compton resigned in 1953 and Ethan Shepley became acting chancellor. Within a year, Donald A. Fisher acceded Stout. In a 1955 letter to Evans, Frank Bubb wrote, "Compton's refusal to make him (Roy Glasgow) dean, and the most unfortunate appointment of Stout, so disgusted me that this was one of my reasons for clearing out."

Aerial view of downtown St. Louis, with
County Courthouse in center foreground

refine new ideas. For Arline, it meant separation from family and friends. She and her dad ("Pop") would exchange multipage letters in longhand for the next three decades.

In the second week of August 1948, Walter took a ten-day break from teaching. Upon his arrival in St. Louis, he and Arline packed their belongings into their 1947 Oldsmobile and drove west on Route 66. Their journey passed by Sullivan, Missouri, near the Evans family farm.

Walter's 1947 Oldsmobile

Arline remembered the trip as an ordeal. Walter drove at night and attempted to sleep during the heat of the day. The road across the Southwest was a two-lane undivided highway with blinding lights from oncoming traffic. When Walter got behind a slow-moving truck, he would be itching to pass. The road, however, followed the rolling contours of the desert—it was treacherous to pass, especially at night. (Note from author: No wonder none of our family vacations ever returned to America's great Southwest!)

Walter and his family were among tens of thousands of Americans who relocated to the Southwest in the 1940s. It was a decade in which the state of California saw its population increase by 50 percent. His brother Sam would accept an offer from Shell Oil. At first, Sam moved his family to Connecticut, but eventually made his way to Houston, Texas.

A virtual army of recent engineering graduates, many like Walter from Midwestern cities, flocked to Southern California. Having benefited in their youth from Midwestern values and education emphasizing practical engineering principles, their expertise and creativity played a pivotal role in shaping the trajectory of American aerospace innovation.

Eureka (August 1948)

The story of Archimedes' "Eureka!" moment is one of the most famous anecdotes in the history of science. Tasked with determining whether a king's crown was made of pure gold without melting it down, Archimedes pondered the problem tirelessly.

One day, as he stepped into a full bathtub, he noticed the water level rise and suddenly realized the principle of water displacement. Overcome with excitement, he leapt from the tub and ran through the streets of Syracuse naked, shouting, "Eureka! I have found it!"

Archimedes' Eureka moment – I have found it!

Walter had his own Eureka moment, though it did not involve bathtubs or public nudity. His breakthrough in control systems—the birth of root locus—emerged in a classroom, prompted by a simple but profound question. What began as an extension of Paul Profos's flux plotting and the dominant methods of servomechanism analysis soon evolved into a revolutionary way of visualizing control system stability. Like Archimedes, Evans had stumbled upon an insight that changed the field forever.

Technical Genesis of the Root-Locus Method

After the Evans family completed their 2,000-mile journey from St. Louis to Santa Monica—a short drive to NAA's Aerophysics Laboratory on Aviation Boulevard in Inglewood—Walter resumed teaching.

His coworkers preferred "Walt" to "Walter" and so from here on he will be Walt Evans—unless Arline is in the picture; he would always be Walter to her.

According to his course notes, his "root locus Eureka moment" may have occurred as early as Monday, August 23, 1948. On that day, he presented several methods for servomechanism design. His course notes reflect his sensitivity to diverse student preferences that would depend upon the student's particular engineering experience and, hence, their expertise. His goal was to tailor instruction to the needs of each student. His notes read:

> *The choice [referring to servomechanism design methods] depends not only on the problem at hand but also upon the previous experience of the designer. Many engineers accustomed to thinking of servomechanisms in terms of examples they have seen will probably lean heavily on their own physical intuition. For them, the output lag behind the input can be decreased by using "anticipation."*
>
> *Other engineers, particularly those with extensive backgrounds in electric circuit analysis, are apt to prefer the frequency response method. In this method, the servo is described in terms of its response to a wide range of frequencies, and corrective action is taken in terms of band-elimination filters.*
>
> *Finally, a person with extensive experience in differential equations will tend to stay with this method, devoting considerable effort to finding the roots of equations. Unfortunately, no single method alone is as useful as knowledge of all three. Any attempt to classify the advantages and disadvantages of each method is bound to be highly subjective.* [20]

However, the precise date of the pivotal student-instructor exchange is unknowable. It may have occurred as late as September 7, when the scheduled topic was "Graphical Solution for Roots of a Simple Quadratic System."

The Slightly Cubic Problem Led to His Eureka Moment

In her 1974 senior report, "Study of the Spirule," at California Polytechnic State University, Walt's daughter Nancy provided the following expanded account:

> *In the summer of 1948 he was teaching a servomechanism course at North American Aviation in which all possible ways of solving a simple quadratic system had been presented. One of the students then asked how these results changed when the system became "slightly cubic." "Slightly cubic" refers to a servo in which the amplifier delay in developing voltage for the motor is small but not negligible.*
>
> *(N)either the Nyquist diagram [n]or the Bode plot would show the effect of the amplifier time constant on the frequency response of the system. One of the quadratic descriptions, however, had been a plot of the roots in the s-plane. The extra time constant was added to this plot and the roots shifted. This description was instantly preferred by most students in the class. This was the beginning of the root locus method for finding the roots of servo equations.* [21]

Walt proposed a graphical root location technique—what he called the "locus of roots" method. The technique visually represented how pole locations evolved as a parameter (usually gain) varied. This visual and parametric approach, to become known as the root-locus method, offered immediate pedagogical and analytical value. His students grasped it quickly and applied it readily.

On September 14, Walt held his classroom session. His notes are instructive, revealing his application of two lessons learned from his Washington University professor Roy Glasgow. First, referring to a point on the s-plane which lies on the locus (i.e., is a root for some value of gain K), he wrote, "the 'p' point is just a guess that would have been right for the quadratic in which the second time delay was neglected." He went on: "The above procedure represents a trick which is often handy: simplify the problem down to one in which the answer is known. Then change the answer slightly to allow for the extra complications. Another trick which is often handy: Take extreme cases."

MIT, Draper, and Frequency Response Orthodoxy

At the time, MIT's Servomechanism Laboratory, under Gordon Brown and Charles Draper, was the intellectual center of the field. The MIT Radiation Lab's wartime work had culminated in *Principles of Servomechanisms* (Brown and Campbell, Wiley, 1948), which emphasized frequency domain

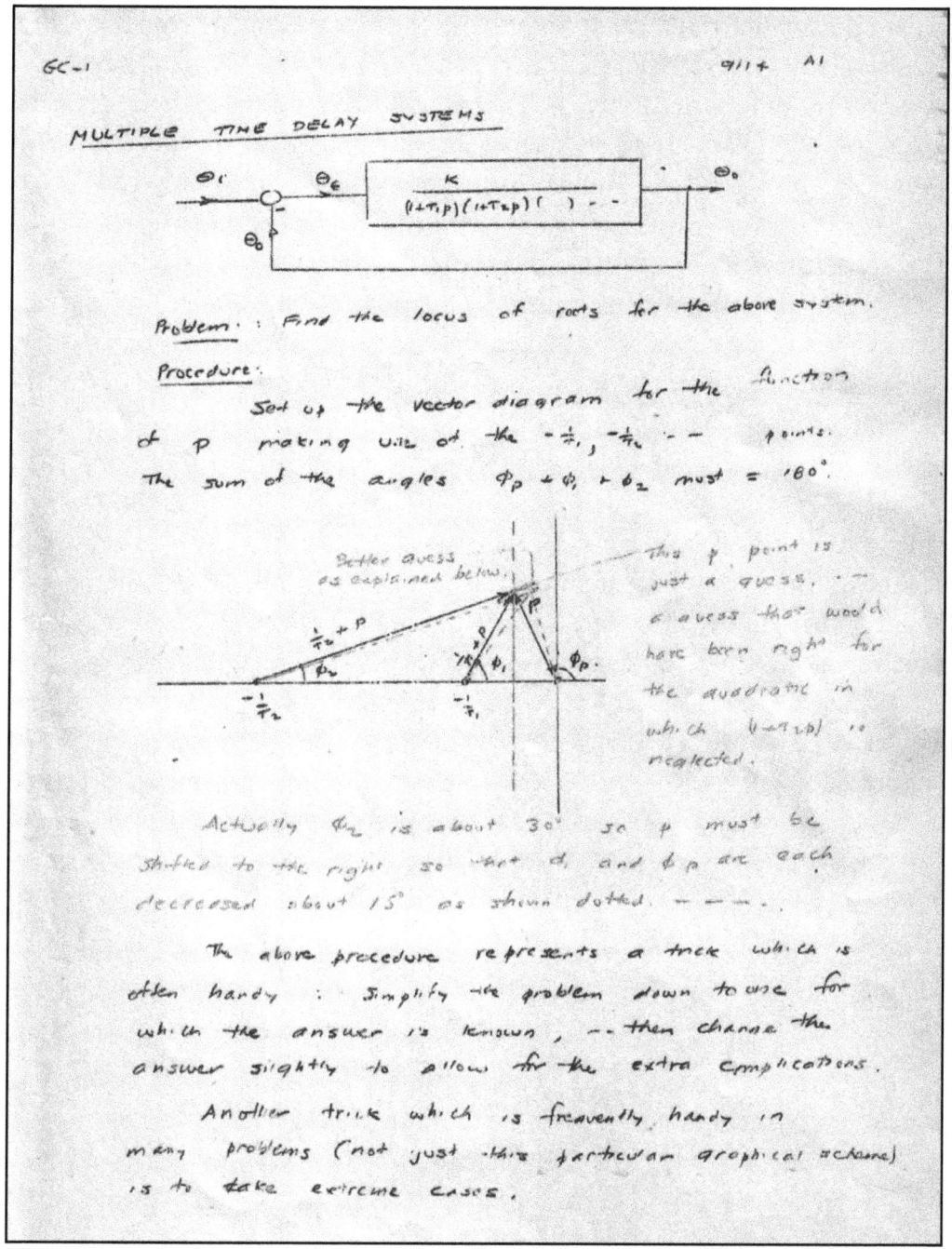

First page of Evans's 9/14/1948 course notes

techniques using Bode plots and Nyquist diagrams. Walt respected this tradition but observed that frequency domain methods did not always yield intuitive insight, particularly in systems with delays, non-minimum phase behavior, or higher-order pole–zero interactions. His root-locus method provided a bridge between analytical rigor and visual intuition.

Publication and Disseminations

In November 1948, Walt submitted a manuscript to *AIEE Transactions*. During the review process, he expanded the concept into a practitioner-oriented internal report, AL-787. NAA printed and distributed it, remarkably, not only within the company but also to engineers at Hughes, Northrop, and faculty at UCLA and Caltech. (AL is for A̲erophysics L̲aboratory.)[22] In today's IP-centric climate, such a contribution would likely be treated as proprietary. But in the postwar period, driven by optimism and collaboration, technical innovation was often viewed as a public good. The willingness of North American Aviation to disseminate root locus exemplified the collaborative ethos of the era—and helped ensure its rapid adoption.

When Walt wrote AL-787 he gave it the title "Linear Servo Analysis by Root Locus Method." However, by March 1949, he replaced the word "analysis" with "synthesis." Perhaps the change was prompted by servo designers like Bill Mullins using root locus to achieve stability by design. The widely circulated, AL 787 was renamed "Control System Synthesis by Root Locus Method."

> **Synopsis of AL-787 March 1949**
> The root locus method determines all the roots of servo equations by a graphical plot, which readily permits synthesis of the system for transient characteristics. The method adopts the block diagram representation of the system now associated with frequency response methods but seeks instead to find the damped sinusoidal signals or real exponential signals that propagate themselves around a loop. The new and key idea is to plot the zeros and poles at the open loop function on the complex plane and then sketch the locus of roots of the closed loop equation as loop gain is increased. The locus is readily sketched because one need only keep the sum of the angles

> on the plot equal to 180°, loop gain can then be selected for a desired damping at a pair of complex roots. For multiple loop systems, one solves the innermost loop first which then allows the next loop to be solved by another root-locus plot. The resultant series of plots shows the effect of each parameter in each loop and suggests the changes or additions involved in synthesis.

Conclusion: The Method as Engineering Paradigm

The root-locus method was not born in a journal, but in a lecture hall. It was crafted in response to a student's question, refined in dialogue with practicing engineers, and strengthened by Walt's pedagogical instincts. What made it transformative was both its novelty and its utility: It provided a graphical, scalable, and intuitive method for assessing system stability and performance as a function of gain.

Walt later reflected on why, of the many solutions he had thought of before their time (e.g., wheels on suitcases, colored tennis balls, staggered work hours, early versions of hyperlinks), the root-locus method was different.

> *I have had more than my share of luck in hitting it big with Root Locus. I mean luck because many other ideas, not directly servo, have failed to arouse any interest. I think that the explanation is that I strive for a kind of understanding that most people don't seek. In the case of root locus, it provided a needed link to a complete solution of a system which many others did seek.*

The root-locus method endured because it filled a gap between theory and practice, and between abstraction and intuition. It became part of the working vocabulary of control engineers because it made invisible dynamics visible. From a question in a summer lecture to a method that would appear in virtually every control systems textbook for the next half-century, root locus stands as a paradigm-shifting contribution.

And it began at North American Aviation in the summer of 1948.

Cardinal by Walter R. Evans, dated June 8, 1985, Portfolio #132, Vol. 2

PART II
FEEDBACK

On careful reading, one realizes that, at the time of publication, Evans's understanding of his method and his ability to use it was at a level that most of us did not reach for another 15 years.

George Thaler
Professor of Engineering
Naval Postgraduate School

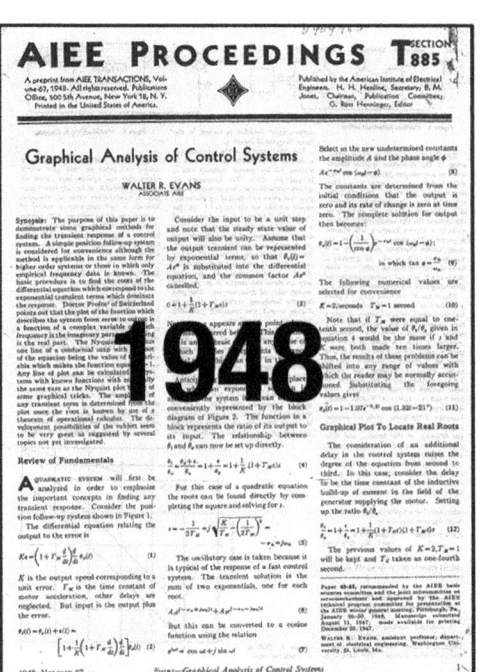

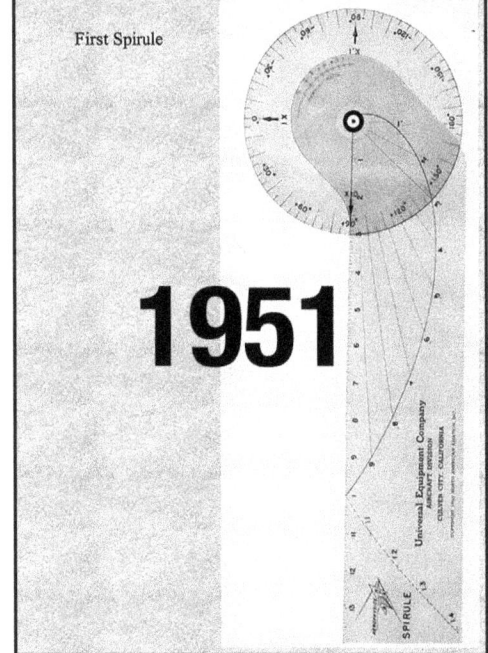
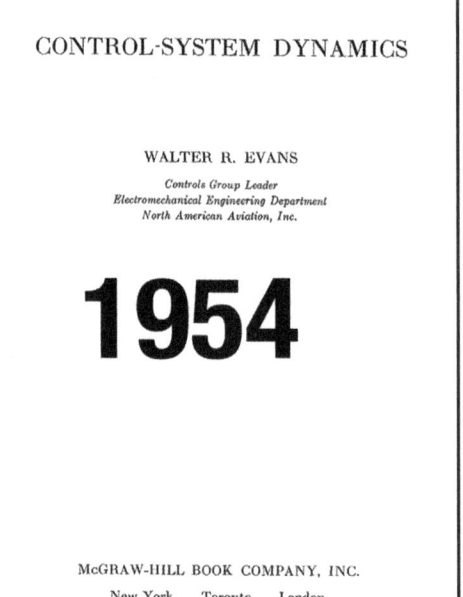

Part II tells the little-known stories behind the better-known achievements of Walt Evans between 1945 and 1954

Chapter 6

Aerophysics Lab (1948–1954)

The Aerophysics Laboratory at North American Aviation became a crucible of innovation, fostering a collaborative environment where physicists, engineers, and technicians tackled complex technological challenges. As one of the first facilities of its kind in the United States, it embodied a broader shift in which private industry assumed roles traditionally held by academia or government laboratories. NAA's strategic foresight in establishing this facility ensured its relevance through the uncertain postwar years and well into the Cold War.

The Navaho Missile and Six Generations: XN-1 to XN-6

The Navaho Missile and Six Generations: XN-1 to XN-6

The biggest contract awarded to the Aerophysics Lab was a powered rocket called the Navaho. During the development of the rocket in the early 1950s, Walt began to ascend the management ladder. "Headed for the vice presidency?" his Washington University professor Roy Glasgow might have asked. Recall that Glasgow had steered Walt away from his declared major of engineering administration. Fueling America's aerospace industry in the 1950s was the threat posed by the Soviet Union. The US and USSR were in an arms race. Schoolchildren were subjected to duck-and-cover drills, in fear not of a shooter, but of an atom bomb.

Dutch Kindelberger, the founding president of NAA, hired J. Lee Atwood as chief engineer in 1934. Atwood eventually became chief executive officer. In a 1989 interview for an oral history project, Atwood described NAA's postwar challenge. He noted that during World War II, the company peaked as a military aircraft manufacturing powerhouse, employing over 100,000 workers nationwide. However, at the war's end, production ceased abruptly, reducing the workforce to about five thousand employees. Atwood explained NAA's strategy for survival.

> *When the [US-built] atomic bomb exploded [in August 1945] and the obvious connection with the possible missiles work and all that became fairly apparent to us, we began to realize that there was going to be a need for a considerable national defense. The measures we took immediately after the war were to bring back everybody we had sent to these inland plants and to push some things we'd started in engineering. By 1948, we had 18,000 people. We were going to do missile work.*[24]

The military's goal was to develop an inventory of bombs. They assumed that each bomb would weigh five thousand pounds. They would require a new unmanned delivery system capable of carrying a payload that heavy more than five thousand miles from launch sites in the continental United States to Soviet targets, with a precision within one mile of the intended target. Achieving these objectives demanded technological breakthroughs in rocketry and guidance systems.

> *I think we moved much faster and much stronger than other companies who had more conventional work lined up. ... We hired quite*

a number of scientists. This was started as early as 1946–47. Dr. Bill Bollay was hired; he was the leader. We set him up with a department—what we called an Aerophysics Department—and that was given a kind of license to explore scientific advancements. … We had Niels Edlefsen in electronics. It wasn't too long before we had quite a stable of well-qualified scientific people, most of them with Ph.D.'s in their fields. This began to grow because nobody else in the airplane business was looking at things quite that way. Bollay initiated the inertial guidance program during this time.

At the same time, of course, the gyroscope authority was Stark Draper of MIT, at the Instrumentation Lab. He'd been working that before the war, during the war, and after, and our Autonetics Division (sic) was not exactly a competitor and not exactly a spinoff, but it was starting to parallel what was going on at the MIT lab and developed a guidance system for the Navaho. There wasn't anybody in the industrial sense prepared to develop that guidance system. And so, we undertook it. The guidance system was tested aboard an old Army transport plane, C-97 (sic), one that the Army could afford to direct to us for test purposes.

Here it is appropriate to break into Atwood's account and explain why Walt's contribution to the design of servomechanisms using the root-locus method was critical to achieving NAA's strategic goal of becoming the aerospace industry's leader in the production of high-accuracy inertial guidance systems. They use servomechanisms to achieve precise and responsive control of its components, which is essential for accurate navigation and stability.

Inertial Guidance Systems Use Servomechanisms

Inertial guidance systems rely on gyroscopes and accelerometers to measure angular velocity and linear acceleration. These sensors must be oriented and stabilized accurately to maintain the integrity of the measurements. Servomechanisms

22-lb Gyroscope for XN-1

> ensure precise control of these sensors, adjusting their positions in response to external forces or changes in orientation.
>
> Servomechanisms are used to counteract vibrations dynamically, keeping the sensors aligned. They use feedback loops to continuously monitor the system's performance and make real-time adjustments. This feedback ensures that the system remains accurate over time, compensating for drift or deviations caused by environmental factors. Servomechanisms allow the inertial guidance system to respond dynamically and maintain accuracy, ensuring the navigation solution remains reliable.

Johnny Moore's "Cream of the Crop"

Dr. James R. Burnett, an executive at Ramo Wooldridge, was envious of the talent Johnny Moore had recruited to work in the Aerophysics Lab. In an oral history recorded in his Thompson Ramo Wooldridge office on June 19, 1989, with Martin Collins, Burnett remarked:

> *Johnny Moore at one time had really the cream of the crop of control system engineers in the country. He had the guy who invented the Spirule—I'll think of his name in a minute—which is a handy tool for determining control system stability. He had Kochenberger over there for a while, who figured out how to do the stability of nonlinear control systems. I mean he had really a bright bunch of guys. Of course, Johnny's very bright himself."* (Emphasis added.)[25]

Walt Evans was not the only rising star recruited by John R. Moore, the consummate recruiter. When one examines the names on Moore's Group 63 bullpen seating chart, dated March 30, 1949, President John F. Kennedy's statement in 1962 at a White House dinner honoring Nobel Prize winners comes to mind: "I think this is the most extraordinary collection of talent …"

The group contains the names of forty young engineers, several of whom were destined to become CEOs. This small group was the envy of executives at other aerospace companies in addition to Ramo Wooldridge. So much concentrated talent!

AEROPHYSICS LAB (1948-1954)

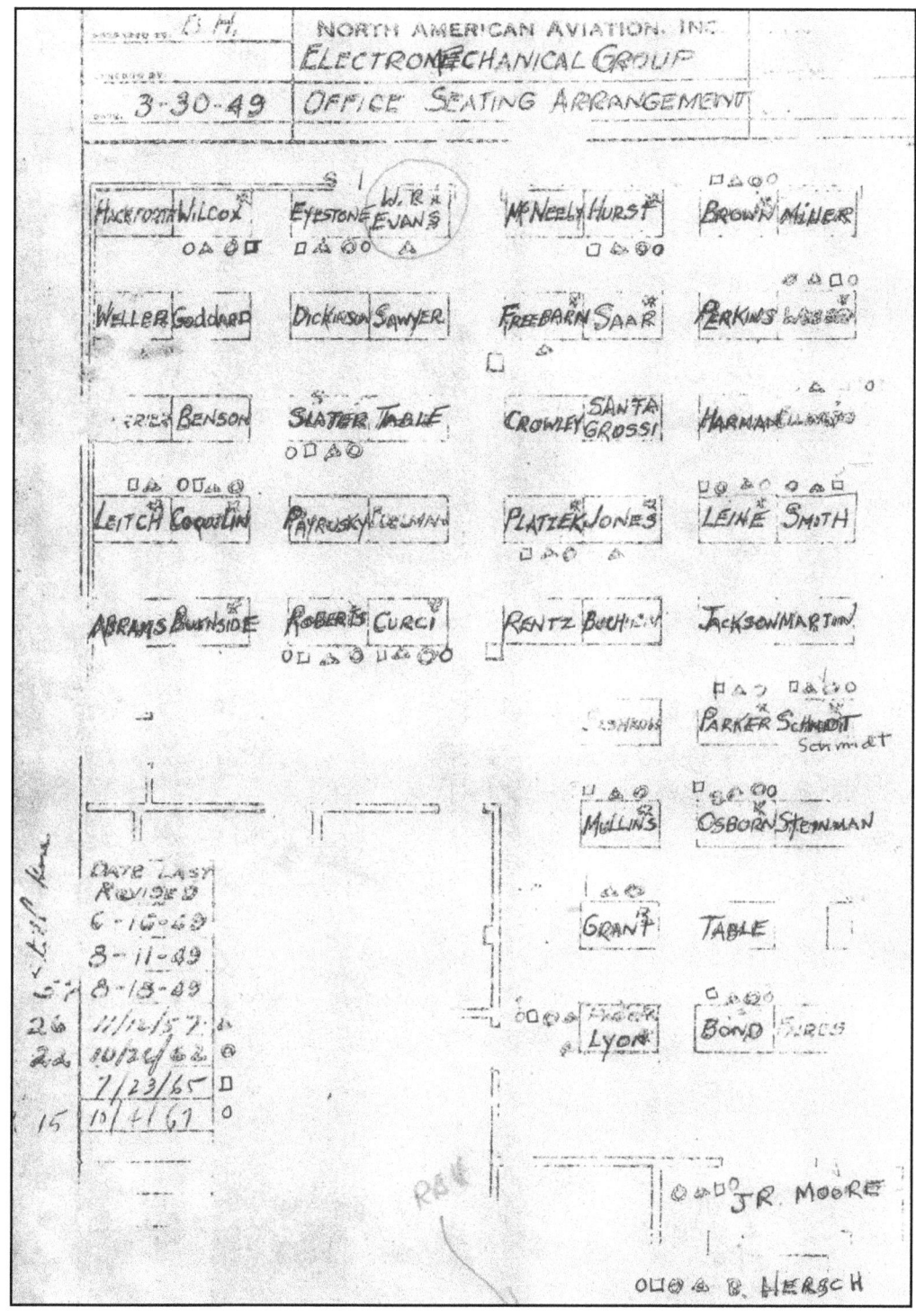

Seating chart in Group 63 Bullpen on March 30, 1949

INTO STABILITY

Moore occupied the executive office cubicle pictured in the lower right-hand corner of the chart. In the previous thirteen months, he had hired many of the forty engineers he could see when he looked across the room. These men became more than work colleagues. Many, like Walt, were recent arrivals to Southern California. Their network of new friends came from

Aerophysics Lab Engineers in John Moore's 1949 Group 63 "Bullpen"

their coworkers. Married employees socialized, forming groups that met regularly for contract bridge, for example. Families met one another at company-sponsored picnics.

The author remembers hearing names from the seating chart at the family dinner table, when Walt talked about his day at work. He sat next to his future boss, **Fred Eyestone**, a few desks away from two excellent engineers—**Doyle Wilcox** and **Bill Goddard**. Goddard would later receive one of IBM's biggest awards for his patent on floating disk drives. **John Slater** led the development of the XN-1's Gyroscope; **Jeff Schmidt** built the first Spirule.

Bill Mullins was the most enthusiastic user of the root-locus method. (In 1951 he tutored Dr. William Bollay when he prepared for his Wright Brothers talk. Walt believed Bollay put the method on the map.) **Al Grant** and **Robert Osborn** sat near Moore. Osborn would play a lead role in administrative decisions about Spirules. **DeWitt Lyon** became a Christian missionary to Japan in 1959, but his family visited the Evanses whenever the Lyons were stateside in the 1960s.

Ward Harman was another good friend who later would promote Walt's ideas at Stanford. Like Eyestone and Moore, **Norm Parker** became a company president.

The novel, bleeding-edge technology they were developing blurred the lines between industry and academia. NAA employees taught night classes at UCLA. Seminars were attended and led by NAA engineers. It was a very special era.

North American Aviation's First Auto-Navigator: The XN-1

The first application of the root-locus method was the XN-1 autonavigator. DeWitt Lyon wrote the official version of the 1950 XN-1 flight test program in the book *Twenty Years of Inertial Navigation at North American Aviation*. He described for the author in private correspondence what *really* happened. The chapter concludes with John Moore's only reference to Walt's role at NAA in his twenty-two-page report "Legacies of North American Aviation Experts." Here are excerpts from Lyon's official XN-1 historical account:

> *In 1948, J. R. Moore joined NAA and assumed direction of the guidance and control effort. Under his leadership, work on inertial guidance was carried on without a break in the basic philosophy, but at an accelerated pace.*

INTO STABILITY

Components of the XN-1 auto-navigator began to be assembled in September 1949; the system work falling under the direction of S. F. Eyestone and being handled by W. R. Evans, J. Y. Bowman, and W. D. Lyon. The system included the XN-1 platform mentioned above, a gyro speed control system which synchronized the gyro speeds to an airborne chronometer, and an analog computer for position computation, acceleration corrections to the distance meters, and earth-rate torquer currents to the gyros.

C-47

Laboratory tests of this system were made in March and April 1950 and the first flight, installed in a C-47 airplane (Serial No. 065), was made on 3 May 1950. The flights were made [from Downey] east to Arlington, near Riverside, California, and return. These flights were from approximately 40 minutes to 1 hour in duration, covering more than the designed flight time for this type of autonavigator (approximately 30 minutes). Flight testing continued through the Summer and Fall of 1950 with this system. [23]

Two employees onboard the C-47 for the initial test flight of an NAA inertial guidance system were Walt Evans and his friend and colleague DeWitt Lyon. Here is DeWitt Lyon's account of what really happened:

Walt, Jesse Bowman, and I rotated as system engineers with the X-1 in the C-47 [military DC-3] flying out of Downey. Each one would work two shifts, flying with the system in the day followed up by working up the data at night. The next day, the next system engineer would do the same, using the previous day's data as reference, while the previous engineer had the day off.

To the best of our knowledge at the time, that was the first flying inertial-navigation system anywhere. You probably have lots of information on the incident when ... Walt's parachute backpack inadvertently tripped a B+ switch in the gyro power supply. The stabilized platform started to drift, but feedback through synchros made the scopes monitoring the gyros look as if they were still synched in!

Someone in the company, hearing of the way Walt had terminated the world's first flight test of an inertial navigator, commemorated the event with an inscription in concrete near the Downey main strip to document it. In his Legacies history[19] John Moore wrote, "I think that one of the great desecrations of modern times was the removal of the historic inscription in concrete near the Downey main strip, with its deathless lines," which were: *From this point in Euclidean space, Man first attempted to trace, His path through the heavens, And in spite of Walt Evans, Safely returned to this place.*

Norm Parker, Walt Evans, and DeWitt Lyon in C-47 when Evans's parachute hit the B+ switch of the power supply, abruptly terminating the test of the auto navigator.

Despite the mishap, the Aerophysics Lab developed more generations of inertial navigation systems to become the undisputed world leader. The technology, aided by root locus, would serve as the guidance systems for the US Air Force and Navy ballistic missile systems Minuteman and Polaris, which—together with the B-1 bomber force—form the triad of the United States strategic defense systems. But applications of the root-locus method extend well beyond its historic role on national defense systems.

Today, root locus continues to play a critical role in advancing safety, precision, and efficiency across a wide range of industries. The widespread adoption of Walt Evans's Spirule—with more than 100,000 units sold—underscores its versatility far beyond missile guidance systems. Modern commercial aircraft rely on advanced control systems to ensure safety and reliability under unpredictable conditions. Autopilot systems, flight control surfaces, and stability augmentation systems must adjust dynamically to turbulence, mechanical faults, and shifting weight distributions.

The exponential growth of drones and autonomous vehicles has brought feedback control to the forefront of technological innovation. Whether delivering packages or conducting aerial surveys, drones must maintain precise altitude, speed, and directional stability. Autonomous vehicles require similar precision for speed control, steering, and braking. Root locus helps optimize these systems, ensuring they remain safe under a wide range of conditions.

In factories and production lines, where precision and efficiency dictate profitability, root locus continues to be invaluable. Industrial robots, conveyor systems, and temperature control processes all rely on well-tuned feedback mechanisms. Root locus helps engineers fine-tune these systems.

Life-supporting medical devices must operate with extreme precision to safeguard patient health. Ventilators, insulin pumps, and cardiac monitors use feedback control systems to maintain critical physiological parameters within safe ranges. Root-locus techniques are instrumental in designing these systems.

As the world shifts toward renewable energy, maintaining grid stability becomes increasingly challenging. Wind turbines and smart grids must manage fluctuations in power supply while meeting the continuous demand for electricity. Root locus helps engineers design controllers that optimize stability and efficiency, even as conditions change.

The continued relevance of root locus across diverse industries reflects not only the brilliance of Walt Evans's method, but also the enduring need for well-tuned control systems in an increasingly automated world. The next time a commercial flight lands safely or a drone delivers a package with pinpoint accuracy, it is worth remembering that root locus helped make it possible.[33]

And Dr. Burnett, the name you could not remember was Evans—Walt Evans.

Chapter 7

The Spirule (1948–1952)

Joined at the hip with the birth of the root-locus method was the need for a tool to help designers sum angles and multiply distances from a trial point "p" on the s-plane to other fixed points. Walt's 1948 classroom notes capture his early attempt to meet that need:

> *Place a transparent sheet on top of the diagram, draw the red line for reference, stick a thumbtack through at the guess point to serve as a pivot.*

Engineers rejected his makeshift solution. They wanted something better—something precise, durable, and repeatable. Yet few at the time, including Walt himself, could have anticipated how dramatically this problem would shape the next stage of his career. The tool would eventually be called the "Spirule." Its journey from plexiglass prototype to globally distributed teaching aid was a story no less remarkable than the method it served.

Jeff Schmidt, Inventor (1948)

Jeff Schmidt was one of Walt's colleagues. His assignment in the summer of 1948 was to design an analog autopilot for the NATIV missile. It had to work without adjustment throughout the changing mass and center of gravity during the launch phase. Schmidt was stymied until Walt presented root locus. In 2003, Schmidt recalled:

INTO STABILITY

Jeff Schmidt, an Aerophysics Lab colleague of Walter Evans, was the first to have the idea of using a plastic circular protractor to aid in creating root-locus plots. Mr. Schmidt prepared the sketch of the device he first made in 1948 below in December 2003. Another Aerophysics Lab engineer, DeWitt Lyon, coined the name "Spirule."

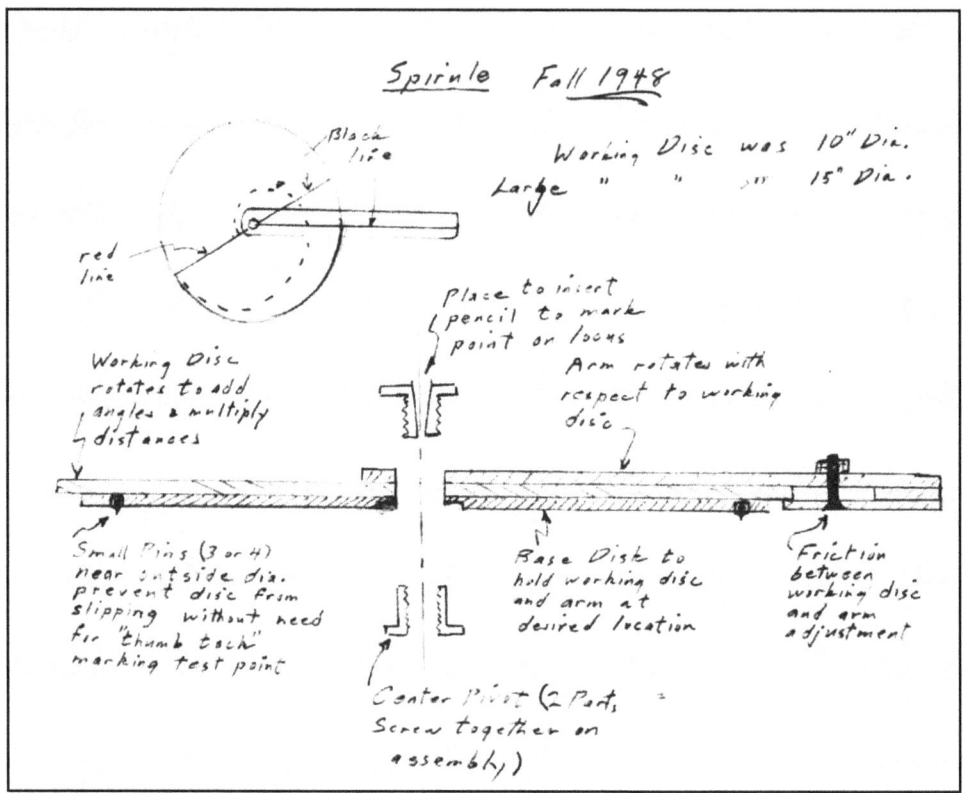

Jeff Schmidt's 2003 drawing of his 1948 Spirule

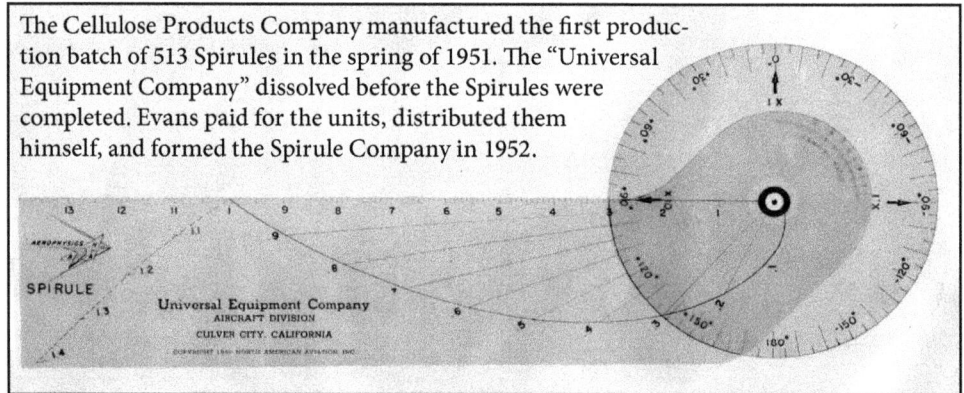

Spring of 1951, First manufacturing run of 513 Spirules

THE SPIRULE (1948-1952)

When Walt first started discussing root-locus in his class, I saw a ray of hope. While the problem was still very difficult, I could at least get a feel for what I was trying. As more people started to use Walt's method, many ideas for automating the plot were thought of. I was looking for something better than [Walt's] transparent paper to add angles. I went over to the engineering shop and made an "angle adder" out of a circle of plexiglass with a straight arm held on with a small bolt. This worked much better than the transparent paper for determining the locus, but I still had to measure lengths and multiply them to get loop gains. Walt and I were kicking this problem around one day, and we came up with the idea of adding a logarithmic spiral to my angle adder. This worked well and became the first Spirule. I applied for a patent, but the company decided a copyright was more appropriate. (See Appendix 5 for Jeff Schmidt's account.)

The Name: The spiral-shaped curve was key to simultaneously using the principles of a linear slide rule (i.e., multiplication by addition of logarithms) and performing the addition with the circular protractor. DeWitt Lyon, a colleague of Schmidt and Walt, looked at the spiral curve on Schmidt's device and coined the name "Spirule" as a contraction of "spiral" and "slide rule."

Ties between UCLA and North American Aviation were particularly strong in those days. Dr. Edlefsen, deputy director under William Bollay of the Aerophysics Laboratory, hired UCLA's Professor John Barnes as an assistant. In September 1948, John Moore became UCLA's first "Visiting Associate Professor of Engineering" and began evenings and Saturdays to teach a graduate-level course on the principles of servomechanisms. Several years later, when the Spirule became available, he required his students to purchase them from the UCLA bookstore, whose purchases over the years would exceed those of any other institutional customer.

In the spring of 1949, Moore invited Schmidt to lecture his graduate servo course at UCLA on the use of his Spirule. Schmidt recalled in 2003:

After the first Spirule had shown its worth, a design was made for a more precise model. A bunch were made for use in the department. After my demonstration in John Moore's servo class, employees of several large aerospace companies asked for a set of drawings. Hughes had several hundred made, but the person doing the assembly looked at the assembly drawing and cemented the parts together in that position. Since nothing could rotate, they were worthless. Walt commented that Howard [Hughes] could afford to build another set but to advise him not to try making them of wood like his Spruce Goose.

Among the students in Moore's UCLA class was a Northrop engineer, Duane McRauer. The Northrop Corporation produced a "root-locus plotter" in the spring of 1949. On the backside of the drawing is the following handwritten note:

Dear Walt, This is a print of the root-locus plotter dreamed up here at Northrop. I believe it will solve all problems of friction, etc. We are having several made up by Vard in Pasadena. Best regards to you and the boys at Aerophysics. Richard Kulda

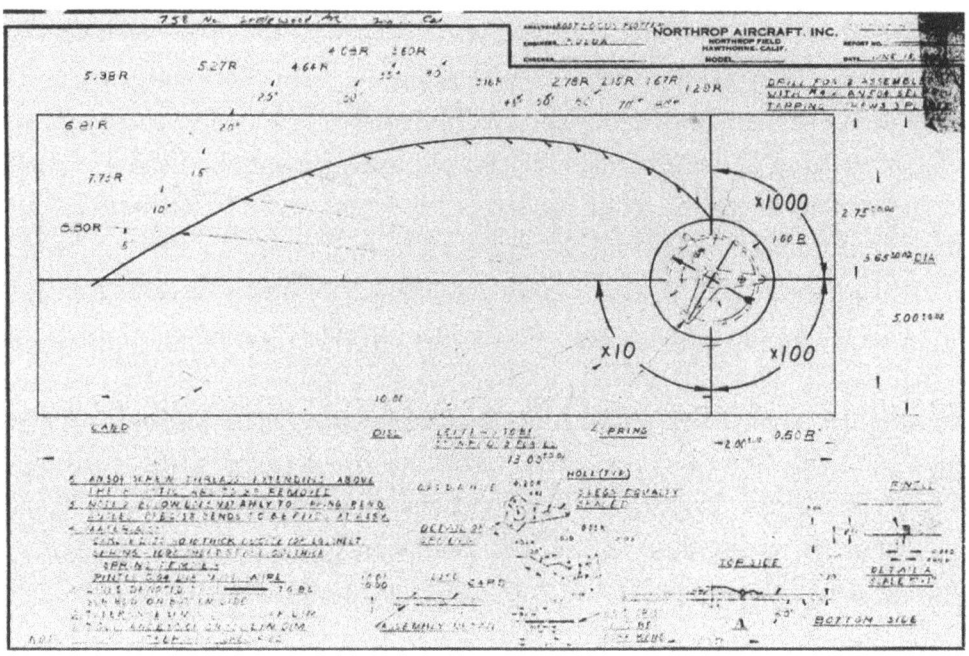

Drawing of Northrop's Root-Locus Plotter

The enclosure dated June 15, 1949, contains a drawing of a 13 in. x 5 in. Lucite card with a spiral curve and a one-inch diameter inset disk affixed to the card with three radial spring-loaded legs. A 0.03-in. "pointed fintle" at the center of the disk provided the pivot point. The drawing shows a rather smallish disk without any markings that would permit it to be used to add angles. It does have the characteristic spiral curve on the arm radiating outward from the pivot point. Kulda's attention to "the problems of friction, etc.," along with the detail the drawing gives to the construction of the attachment of the disk and the arm at their common pivot point, foreshadowed things to come. No other aspect of future Spirule construction would receive nearly as much attention as its pivot mechanism.

Walt's records do not reveal whether Kulda ever had his design constructed by Vard. The name Richard Kulda is absent from all other correspondence. A Google search found a record of the Vard Company in Pasadena and its founder Vard Wallace. It was a supplier of drafting equipment and, interestingly, an unusual telescopic front fork Vard built for Harley-Davidson. The mystery of whatever became of the Northrop design remains unsolved.

Spreading the Word: NAA Educates the World About Root Locus (1949)

In 1949, NAA organized a series of "Servo Seminars" at which engineers such as Walt, DeWitt Lyon, Norm Parker, Ken Jackson, R. M. Osborn, and John Moore taught their colleagues in their respective areas of expertise. Interestingly, Jackson and Osborn had already mastered root locus well enough that they made the presentations on its use in the design and analysis of autopilots. Walt met with engineers of all stripes—young and old, impressionable and stuck in their ways. Some went away simply impressed by Walt's mastery of the concepts. Others followed and themselves became effective and enthusiastic champions of root locus. These people included men like Ward Harman, Ken Jackson, Bob Cannon, Jeff Schmidt, Mark Campbell, Bill Mullins, and John Moore. It was through these root locus "disciples" that the ideas began to spread, even as the AIEE vetting process dragged out the paper's publication.

Another school teaching servomechanisms to engineers from aerospace companies was Caltech, where a young professor, Charles Wilts, taught the servomechanism class. He incorporated AL-787 into a class he taught at Caltech in 1949. In a letter dated October 1, 1949, Wilts sent Walt a personal letter thanking him for making AL-787 available to him; he enclosed a list of thirty-five names of engineers affiliated with ten companies and government organizations who wanted their own copy. (Almost two years later, in

October 1951, Caltech's Wilts would be among the first to order a Spirule.) In 1967, the author had Professor Wilts for Electromagnetism Theory as a sophomore at Caltech. He had no knowledge of the pivotal role Wilts had played in the early days of root locus. He was an energetic man known to his friends as Charlie.

Feedback from the AIEE Root Locus Paper: Interest in Spirules (1950)

A milestone in the history of the Spirule occurred in January 1950 when Walt's ideas gained their first national platform at the January 1950 meeting in New York of AIEE's Feedback-Control Systems Group of the AIEE (Note: The suspenseful story of events that led up to his presentation is the subject of Chapter 8.).

Walt traveled to New York to present his paper. Surprisingly, the paper's only mention of the Spirule is as a do-it-yourself project: "The reader can duplicate the 'spirule' with two pieces of transparent paper, one for the disk and one for the arm." [27] Hence, many in the audience would have never seen a real Spirule.

During his talk, Walt introduced the Spirule. Afterward, approximately seventy attendees expressed interest in obtaining one. Walt encouraged listeners to duplicate the Spirule *for themselves* with pieces of transparent paper and his words of explanation. This was one of many occasions in which he suggested to others that they need not purchase a Spirule from him to teach or learn the principles involved in root locus—they could create their own. Jeff Schmidt reported that few did. To Walt's surprise, engineers wanted Spirules.

Walt returned to California expecting dozens of follow-up calls. He received exactly three, all in March. The first, dated March 1, came from W. C. Osterbrock, head of the University of Cincinnati's Electrical Engineering Department. On March 21, Professor R. C. H. Wheeler from the US Naval Postgraduate School in Annapolis asked, "Have you made the Spirule for sale? If so, how can I obtain one?"

On March 27, W. E. Meserve, Professor of Engineering at Cornell, reminded Walt that he had promised to provide more information on obtaining the slide rule device.

Walt waited until April 1950 to respond, explaining that he had delayed until he could distribute copies of his AIEE paper. His response to all three correspondents included the following:

Thank you for your interest in the root-locus idea. At the AIEE meeting, approximately seventy attendees expressed interest in obtaining a Spirule. Since then, only three have written. This presents a cost issue: $30 per unit for a machine shop model or 50 cents each for a stamped model in a lot of 500.

The Spirule isn't worth $30, and the demand is far short of 500! It looks like they won't be manufactured until I find an intermediate production method or finish writing a book in which the Spirule will be enclosed.

W. C. Osterbrock replied prophetically, encouraging Walt to manufacture Spirules and make them available for sale. Walt consistently underestimated the interest of engineers in obtaining Spirules.

I wonder whether you are justified in your doubts about the demand for the "Spirule." We could have disposed of sixty-five at a reasonable price, and we shall have another class of about eighty students in servo this summer. Other schools teaching the subject would likely recommend it if properly advertised. I hope you will find a way to produce and make it available.

Manufacturing Challenges (1950)

An October 19, 1950, NAA memorandum from R. M. Osborn to W. J. Toher requested funding to produce five hundred Spirules. Osborn noted that engineers at NAA, Hughes Aircraft, and Northrop had built their own Spirules and that root locus was already part of servomechanism curricula. He estimated production costs at thirty-five cents per unit in a batch of five hundred. He also suggested distributing them as souvenirs to visitors and educational institutions. Despite the appeal, the request became mired in bureaucracy.

Dr. William Bollay's Lecture (December 1950)

On December 8, 1950, Walt updated Dr. Edlefsen:

> *A plastic Spirule sample was made but required one man-day to fabricate, making mass production impractical. We explored photographic printing on a thin plastic sheet with lamination for durability, using a simple eyelet as a pivot.*

One week later, on December 16, Dr. William Bollay delivered the Wright Brothers Lecture on "Aerodynamic Stability and Automatic Control" at the Institute of Aeronautical Sciences in Washington, D.C. The lecture's publication helped establish root locus in the mainstream of feedback control systems. Bollay, who had studied root locus intensely, including time with NAA's Bill Mullins, sent Walt a copy, writing:

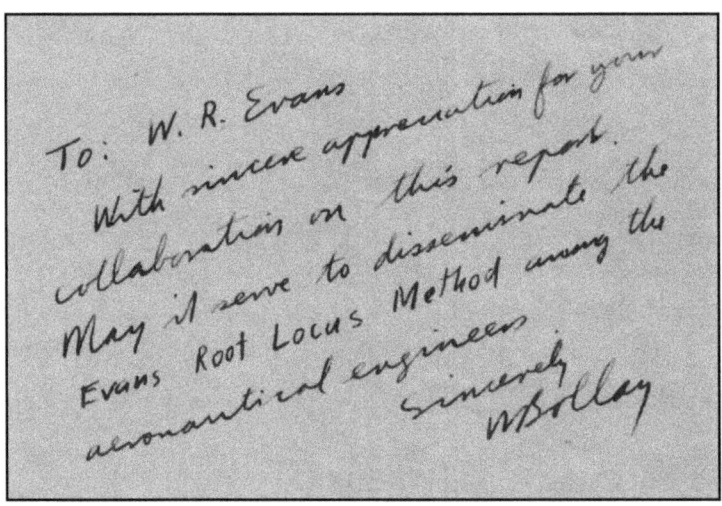

To W. R. Evans, With sincere appreciation for your collaboration on this report. May it serve to disseminate the Evans Root Locus Method among aeronautical engineers. Sincerely, W. Bollay [26]

Spirule Production and Initial Sales (1951)

Meanwhile, interest in root locus, fueled by word of mouth, kept the idea and method alive, even without Spirules. Former NAA colleague Ward Harman had left NAA for a position at Stanford; he enthusiastically promoted root locus. Joseph Chadwick of Stanford's Electronics Research Laboratory wrote on January 8, 1951:

THE SPIRULE (1948-1952)

Ward Harman has been such a convincing salesman of your root-locus method that he's exceeded his ability to inform. Could I obtain a couple of your celluloid gadgets for summing angles, etc.?

And yet, at the dawn of 1951, funding for Spirule fabrication remained elusive.

"Our efforts to produce Spirules have been snarled in accounting procedures because the logical approach is to manufacture them in bulk and distribute them." Walt sought manufacturers, with three companies rejecting the project and three others not responding. Finally, Cellulose Products in South Gate, California, agreed to fabricate the Spirule. NAA contracted with Universal Equipment Company in Culver City, which received the first bill for 513 Spirules on April 30, 1951. When the partnership dissolved, Walt personally paid the bill and all Spirule orders were routed directly to him.

On June 14, 1951, Walt responded to an NAA colleague: "Enclosed is the 'Spirule' long promised to you. I believe you will find it self-explanatory; the instruction sheet is not yet prepared." Walt proceeded to distribute thirty-one Spirules to other colleagues, including Bill Bollay, John Moore, Norm Parker, Jeff Schmidt, and Walt Pondrom, keeping a list at his desk.

Following the September 1951 publication of Bollay's lecture, Spirule sales officially began. On October 2, an internal NAA form recorded the first paid orders. The first academic orders arrived from Stanford, Caltech, and MIT, while Bell Aircraft placed the first corporate order. Orders continued from institutions such as the University of Illinois, Washington University in St. Louis, the University of Washington, and the University of California.

By the end of 1951, root locus and the Spirule had firmly established themselves within the engineering community. Growing interest in the root-locus method (and, of course, the Spirule) from engineering school department heads and professors led textbook publishers, especially McGraw-Hill, to explore a new series of servomechanism textbooks.

The Spirule

A plastic device with a disk and an arm with a common pivot point.

The disk and arm are held together with a light friction fit by an eyelet; a small pin is held in the center of the eyelet by a clear plastic plug.

The pin is stuck in the vector plot at the desired s point to form a pivot for all measurements at the point. Note that all phase angles and vector lengths on the plot have this s point in common.

The total phase angle is obtained by rotating the arm with respect to the disk through each of the phase angles in succession; the total angle is read on the disk at the radial edge of the arm.

The spiral curve on the arm is plotted such that the angle V from the radial edge to the curve is proportional to the logarithm of the radius to that point on the curve. The arm is rotated through each of these angles in succession in order to add logarithms.

The numerical value of the product is read on the curve in line with an arrow on the disk subject to the correction xn, in which x is the numerical value on the plot corresponding to 5" and n is the excess of poles over zeros. Any such correction factor can be set into the Spirule initially and marked on the frosted surface so that all further calculations give the final value directly.

Terms in F(s) of the form (1+Ts) must be treated as a ratio (1/T + s)/(1/T). The factors such as 1/T can be set in by first pivoting the Spirule at the origin and rotating the arm opposite to the usual sense.

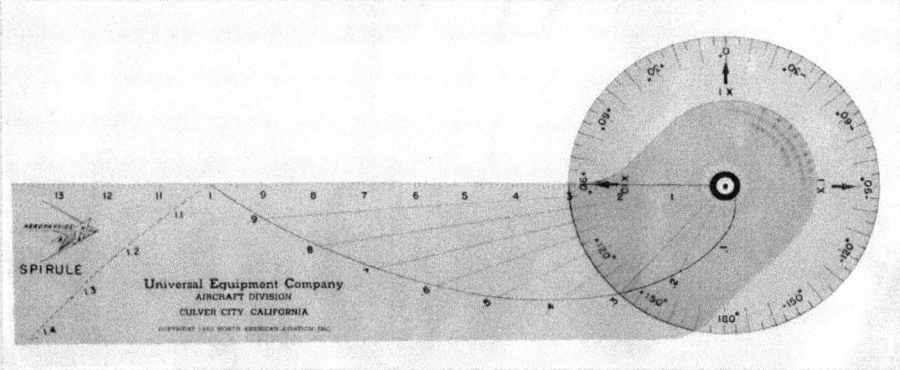

First institutional order in October 1951 from Caltech for two Spirules

College Bookstore Orders (January–March 1952)

Orders continued to flow in, some sent directly to Walt at his Whittier home, some to him at his NAA Downey address, and some to the Universal Equipment Company in Culver City. The most significant development in the first quarter of 1952 was the enthusiasm of two engineering professors from the University of California—Joseph Beggs of UCLA and Otto Smith at UC Berkeley.

The UC schools' embrace of root locus and the Spirule into their departments' feedback control classes laid the foundation for Spirule sales to university bookstores. Joe Beggs and Otto Smith were in a vanguard of professors at UCLA, Berkeley, Stanford, and Caltech who chose to introduce the new root-locus methods into their classes. Many of these used Brown and Campbell's 1948 *Principles of Servomechanisms* as their text. The March 8 order for twelve Spirules from Berkeley's associated student store notes:

> *We have been asked by Prof. Otto Smith of Electrical Engineering to carry these in our department. We do not know the rule but he states that there will be a continued demand for them. This rule was invented by W. R. Evans.*

Berkeley's bookstore seemed confused about how to order their Spirules. Walt received their first order via the "Robison Corporation" at Universal Equipment's Culver City address. Their second order was addressed to the "W. E. Evans Company" (sic) at Walt's home address on Maple Street.

1706 Maple Street

In March, orders were received for fifty-five Spirules for resale by UCLA bookstore, thirty for resale by the University of California at Berkeley bookstore, and twenty for Charles Wilts's servomechanism class at Caltech. The supply from the initial batch of five hundred had begun to run dry. Even worse, the Universal Equipment Company partnership had dissolved; the new owner wanted to raise the price of the Spirule. Walt worried about the affordability of the Spirule for students. Having successfully completed the assembly, correspondence, billing, and shipping himself for the entire manufacturing run of five hundred units, Walt offered and was granted permission to assume full responsibility in the future.

That story is told in Chapter 11. But first come Chapters 8, 9, and 10. They tell the story of the root-locus method's publication in a landmark journal article by the AIEE, its publication in a textbook by McGraw-Hill, and its application in an inertial navigation system by NAA that guided a submarine to the North Pole.

Chapter 8

The Root Locus Paper (1948–1950)

Saturday, May 7, 1949

Occasionally history delivers a coincidence so startling, it defies explanation. July 4, 1826, was such a date. Fifty years to the day following the signing of the Declaration of Independence, two giants of the Revolutionary War, John Adams and Thomas Jefferson—at times bitter political rivals but by then reconciled—both passed into history within hours of each other, one in Massachusetts and one in Virginia. Saturday, May 7, 1949, may fall short of that standard. However, as the AIEE review committee extended its six-month resistance to publication, the coincidence that did occur highlights the significance of what Walt believed was at stake.

The Soviet Union and the United States were in an arms race in an undeclared war. Unlike the World War recently ended, this new "Cold War" would not be led by generals or fought by soldiers on battlefields. Rather, it would be led by scientists and waged by engineers in their laboratories and rocket test ranges. All the engineers working in Johnny Moore's bullpen knew the stakes.

Saturday, May 7, USA—1706 Maple Street, Whittier, California

Despite Saturday being a day off, Walt was angry. At the Polo Grounds, the New York Giants had delivered a 9–1 drubbing to the St. Louis

Cardinals—his hometown team. But his frustration had less to do with the loss of one game and more to do with six months of waiting.

In December 1948, he had submitted his second servomechanisms paper to the American Institute of Electrical Engineering (AIEE). As in 1947, he found himself caught in the agonizingly slow review process. He had already missed the 1949 winter general meeting because one critic complained about the paper's length. Now, two months after submitting a shorter version, he was still in the dark about its acceptability.

Would it take a full year, just as it had with his 1947 graphical analysis paper, before his new root-locus idea would get a hearing at a general meeting? Were open-minded reviewers assessing his work, or was his paper in the hands of close-minded critics? At work, younger engineers embraced his new method.

Dear Professor Brown (First Attempt)

Frustrated, Walt sat down on May 7 to vent his anger toward the one person he believed controlled the fate of his paper: Professor Gordon S. Brown, founder and director of MIT's Servomechanism Laboratory and chairman of the AIEE Technical Program Committee's servomechanism subcommittee.

> *The AIEE has now had the "Root Locus" paper for nearly <u>six months</u>. … In my humble opinion, this is a pathetic record for an Institute that claims the "dissemination of <u>new</u> theories" as one of its fundamental aims.*
>
> *This is a subject on which I might write in undeniably clear language if there were reason to believe that it might do some good. …*

He finished the draft of his letter and set it aside.

Saturday, May 7, USSR—Kapustin Yar Launch Complex Near Volgograd

On this day, the Soviets carried out their first test flight of the P-1A rocket. It was a success. A variant of the R1-A rocket, this new version was equipped with a detachable warhead. The R1-A "Scunner" rocket was a version of the German V-2 rocket that had been used with devastating effect in World War II. The Soviet engineers had apparently overcome their most vexing problem—maintaining stable flight control—to achieve a successful 270-km flight. Five more successes would follow in the month of May. [34]

If Walt had known of the critical milestone that the Soviet Union had just achieved in their rocketry program, only hours before he had sat down to vent, his anger at the AIEE review committee would have been magnified. Walt had been a key contributor for nearly twelve months in John Moore's Group 63 engineering brain trust (see Chapter 4). Thus, he knew the potential national security implications if defense contractors at other companies remained ignorant of the best available methods for achieving stable flight control of their unmanned warhead delivery systems—whether they were powered rockets like the Navaho or ballistic missiles under development at other contractors.

This Soviet success on May 7 was unknown to Evans, but he was well aware of America's competition with the Soviet Union. Three years had passed since Winston Churchill had famously declared in Fulton, Missouri, "From Stettin in the Baltic to Trieste in the Adriatic, an iron curtain has descended in the continent." By May 1949, the USA–USSR relationship, never without issues, had grown more tense. Less than a year had passed since the Soviet Union tried to block Western access to Berlin, Germany. President Truman had responded with the Berlin Airlift, a vital lifeline for the besieged residents of West Berlin.

Dear Professor Brown (Second Attempt)

Later that evening, Walt returned to his draft. Should he send it? What good would it do? His mind churned. Creative but impatient by nature, he seldom settled on the first solution that came to him. Instead, he sought out alternate perspectives that might reveal a better approach. Walt knew that John Moore and Gordon Brown had met recently and likely discussed his paper. Moore would surely report back soon.

Though we cannot read his mind, we know what Walt did: He rewrote the letter, this time laced with sarcasm and a veiled threat to publish in a different journal.

> *The Root Locus idea ... has been chasing around <u>for six months</u> in some "multiple loop" with no visible "output." The only "feedback" has been rejection and your recent letters. There must be something*

> *"nonlinear" in the system. ... The obvious alternative is to submit the paper to a different organization.*

This second version of the letter went by airmail. On his first draft, he scrawled, "Not sent."

Dear Professor Brown (Third Attempt)

Sunday, May 8, was a better day—at least for the St. Louis Cardinals, who trounced the Brooklyn Dodgers 14–5 at Ebbets Field. The 1949 baseball season was still in its early stages. As it turned out, so was the review of Walt's paper. The Cardinals were recent (1946) world champions, led by 1948 batting champion Stan Musial, and had their sights set on the pennant. To get there, they would have to outlast the Brooklyn Dodgers.

Monday morning, May 9, Walt drove down Imperial Boulevard from his modest home in Whittier to the Aerophysics facility in Downey. There, he learned from John Moore that the March 28 revision of his paper was *still* unacceptable to Brown. Worse, Moore's comments suggested that Brown had not yet grasped the value of the root-locus method.

Anxious to understand exactly what Brown didn't see, Walt wrote a *third* letter before Brown had even received the one from May 7.

> *Dear Prof. Brown, John Moore returned with news that the "Root Locus" paper is still unacceptable to you. ... Therefore, I would still appreciate receiving a copy of the paper marked with your questions.*

He ended with a veiled ultimatum:

> *I see no alternative to starting some action in parallel with your committee. The slow and inconsistent record on the AIEE decisions on this paper and T885 do not justify any further delay. Sincerely yours, Walter R. Evans*

At home that evening, Walt vented further. He took out his manual typewriter and banged out *four* more letters in one night. Each was to men he hoped would be more receptive than Brown. They included his former General Electric colleague, Gordon Walter; his first-year supervisor, Orrin Livingston; his old instructor and now head of the department, Louis (Doc) Rader; and his Washington University contact, Phil Michel.

To Louis Rader, he wrote bluntly:

Remember the low opinion I had of the AIEE technical paper system in Pittsburgh? Well, it has now sunk even lower. The enclosed paper was first submitted to the Winter Meeting. … Every bit of information I've received has been the result of needling them. … The engineers here at North American's Aerophysics Lab have already adopted it over standard methods. If you or any of the boys have questions, I'd be happy to answer them.

Gordon Brown's Response

On Tuesday, May 10, Brown received Walt's May 7 letter. He responded with a carefully worded yet unmistakably stern reply. Taking the unusual step of requesting his secretary send it by airmail, he enumerated his points, five of which began with "please." By Thursday evening, May 12, Walt had the letter in hand. He opened it to find a structured rebuke—polite but sharp. One passage stood out:

Please remember that anyone who seriously undertakes the job of reviewing manuscripts spends a great deal of time in the interests of the author's professional reputation. This work is purely voluntary and part of the code of professional ethics.

The final note advised: "Please … talk to Mr. Moore upon his return to California and get his advice before you let this matter go too far."

At the bottom of the letter, Walt saw that Brown had forwarded copies to three members of AIEE's Technical Committee and John Moore.

Walt was both contrite and angry. He immediately drafted a point-by-point response. To Brown's fourth comment, he typed:

You may excuse my overlooking the time volunteered by reviewers when you realize that 'too long' and 'show contribution to synthesis' were essentially the only reasons given for the initial rejection. The hundreds of hours I spent developing the method were fun, but they have now been matched by write-ups and revisions. More importantly, think of the thousands of man-hours wasted by AIEE members on inefficient methods due to this publication delay.

By the time he finished, it was late. He set the letter aside to sleep on it.

Dear Professor Brown (Fourth Attempt)

Following Gordon Brown's suggestion, Walt sought guidance from John Moore. This time, he learned that Brown had told Moore he had spent hours reviewing the root-locus paper and had even considered rewriting it himself. From colleagues, Walt discovered that their compliments on his paper stemmed partly from their awareness of the time he had devoted to it. His anger dissolved into a mix of contrition and embarrassment.

Setting aside the defiant draft he had written the previous evening, Walt composed a letter of apology for his secretary to type and send to Brown. For the fifth time in six days, he began with:

> *Dear Professor Brown, Nothing seems less funny than an attempt at humor that is out of place. I wish to apologize for my letter of May 7; it was an ill-considered outburst of long-accumulated frustration. … Any effort you can make to ignore my recent letters would be appreciated. I have been completely deflated. The only thing I'm sure of at this time is my duty to revise the paper until it is acceptable.*

Over the next two weeks, Walt did just that. On May 29, he submitted his revised paper to the AIEE in New York, changing the title to "Control System Synthesis by Root Locus Method" and removing the term "servomechanism." He also sent a copy to Gordon Brown at MIT.

The next response, however, came neither from Brown nor the AIEE. Instead, it was from GE's Orrin Livingston, replying to one of the four letters Walt had written on May 9. Livingston confessed his bewilderment:

> *I struggled through a few pages and am now referring to some of the other boys, hoping they can explain it to me in words of one syllable. … Incidentally, if you have anything of a more elementary nature, we'd like to receive a copy.*

By this point, Walt had already made another attempt to clarify the root-locus method. Though Brown controlled the fate of his paper, Livingston's confusion bothered him, and another idea began to take shape.

A week later, a letter from MIT arrived. This time, Brown's comments were significantly more favorable. Walt's apology and revised paper had their intended effect, as Brown wrote:

I am pleased to acknowledge receipt of the carbon copy of the final manuscript of your paper ... I believe you have done an excellent job incorporating the reviewers' suggestions ... I assume you sent the necessary extra copies for review.

Of course, Walt knew better than to assume this was the final step; he had already sent the required copies for further review.

Meanwhile, Livingston's request for a simpler explanation gave Walt another project. The formal tone required for AIEE papers had been a challenge. While his intelligence commanded respect, his colleagues—his wife, John Moore, and others—were drawn to him just as much for his sense of humor. (Years later, for example, he would hire an applicant while playing touch football with him.)

Realizing he had never described root locus informally, Walt used Livingston's letter as an opportunity. By June 13, he had written four handwritten pages titled "The Root Locus Idea" with sections on "Problem, the 'Classical' Approach"; "The Frequency Response Technique"; "Roots—Their Lives and Habitats"; and "The Root-Locus Plot." The only surviving copy remained in his files—he never had it typed or published. In his letter to Livingston, Walt explained:

Actually I've long thought it would be a sporting idea to write up the root locus method the way I came to understand it—free from "approved terminology." I came to understand that it might be a while before getting around to it. It seemed wise to write immediately with another approach. Forget the synopsis and introduction. Let's concentrate on the simple cubic system. That's just tough enough to illustrate the idea and no more. Incidentally, it took from 1946 to 1949 to hit upon the idea, and all the rest was worked up in three weeks, so I'm sure you can work out all the rest anyway. Good luck with the enclosed write-up.

(Walt's June 1949 letter to Orrin Livingston can be found in Appendix 4.)

> "Servo" in "servomechanism" means "slave" in Latin. One example of a servomechanism is the complex machinery that causes the Hubble Telescope to point exactly at a particular galaxy when an astronomer enters the galaxy's coordinates. The servomechanism tries to reduce to zero the difference between the direction the astronomer has told the telescope to point and the direction its sensors tell it is pointed. Machines take time to respond to commands, however, and these delays induce an oscillatory behavior. Servomechanism designers want the amplitude of these oscillations to always decrease over time. This behavior is "stable." Oscillations that grow over time are "unstable." Walt's root-locus method provided a new way for engineers to be confident their designs moved into stability.

Goal: Bring Root Locus to the Classroom with Accessible Mathematics

That summer, Walt set aside any further revisions on hand and turned his attention to writing a book on root locus. The full story of that endeavor is covered in Chapter 9, but it began with a letter to John Wight, Editor of Engineering Books at McGraw-Hill in New York. In it, Walt outlined his vision: "The main purpose of the book is to demonstrate the root locus method."

Meanwhile, global events underscored the importance of his work at the Aerophysics Laboratory. On August 29, news broke that the Soviet Union had detonated its first atomic bomb, "Joe 1," in Kazakhstan. It was a near-exact replica of the American weapon, built with intelligence obtained from sympathizers and spies. Three days later, on September 1, Walt signed a memorandum of agreement with McGraw-Hill to prepare and supply a book titled *Control-System Dynamics*.

Regarding his root locus paper—still under AIEE review—Walt received unexpected support in August from G. D. McCann of Caltech, who wrote to one of the reviewers, Sy Herwald:

> *I believe this is such an important contribution to the art of steady-state analysis of linear systems that every effort should be made to have it accepted by the AIEE as soon as possible.*

In September, the baseball season ended. Although it went down to the wire, the Dodgers edged out the Cardinals for the National League championship by a single game. Stan Musial's 0.338 batting average fell a few points short of the 0.342 posted by the Dodger's new leader, Jackie Robinson, who won the batting championship. Just as it had with his graphical analysis paper, an entire baseball season had come and gone while he had seemingly struck out again with the AIEE.

Finally, in mid-October, Walt received the word he had long been waiting for: The AIEE would recommend his root locus paper for publication. However, due to a packed fall schedule, it would *probably* be presented at the 1950 Winter General Meeting.[27] Which is exactly what happened.

Impactful Root Locus Influencers: John G. Truxal and George J. Thaler

The ideas that Walt finally had the opportunity to present in New York in January 1950 made a very positive impression on one man whose opinion would really matter: MIT graduate student John G. Truxal. In August he published his dissertation, "Servo-mechanism Synthesis Through Pole–Zero Configurations," before joining Purdue University and, in 1954, the Polytechnic Institute of Brooklyn. He would devote three chapters and 190 pages of his 1955 textbook *Automatic Feedback Control System Synthesis* to Walt's method (chapter 4) and its application to system design (chapters 5 and 6), leading to root locus reaching thousands of engineering students.

Another influential advocate was George J. Thaler, destined to become an award-winning teacher and textbook author. Years later, in 1974, Thaler, now a professor of engineering at the Naval Postgraduate School in Monterey, California, compiled and edited *Automatic Control: Classical Linear Theory*, a book of twenty-one classic papers.

The last papers he chose to include were the 1948 "Graphical Analysis of Control Systems" and 1950 "Control System Synthesis by Root Locus Method" papers that Walt had persisted through years of waiting, wrangling, and rewriting from 1947 to 1949 to have published in *AIEE Transactions*. In his introduction of them, Thaler wrote:

> *This group of two papers comprises the only post-World War II contributions in this volume. In these papers Evans introduced the now-famous root-locus method.*

INTO STABILITY

The first of these papers is essentially background, in the sense that it shows the kind of thinking that later led Evans to the root-locus method, and it also shows the elementary form of protractor which later became the Spirule. Some of the conformal mapping ideas are useful and informative.

Our last paper is, of course, an exposition of the root-locus method itself. Little need be said here, except to point out that the paper is very concise yet contains a wealth of ideas.

On careful reading one realizes that, at the time of publication, Evans' understanding of his method and his ability to use it was at a level that most of us did not reach for another fifteen years.

We have terminated this volume with the classic papers by Evans primarily because they mark the last of the major, fundamental contributions to what is commonly known as the classical theory of linear, continuous feedback control systems. There have been many contributions to classical theory since 1950, but such contributions have been expansions, clarifications, and applications of the fundamental ideas. Shortly after 1950 the explosion in technical publications began.

Many of these papers are of major importance, but none seems to be classic in the same sense as the ones presented here. [28]

Evans's persistence paid off.

Chapter 9

The Textbook (1949–1954)

A version of Chapter 9 appeared as "Bringing Root-locus to the Classroom," in IEEE Control Systems Magazine *in December 2004, Vol. 24 No 6.*

In 1948, John Wiley and Sons published *Principles of Servomechanisms* by Gordon Brown and Don Campbell of MIT. It quickly became a landmark text in the field. Meanwhile, McGraw-Hill's distinguished *Electrical and Electronic Engineering Series* had no entry in the rapidly developing field of servomechanisms. Frederick Emmons Terman, consulting editor of the series, had launched the series with his 1937 book *Radio Engineering*. (Terman is known as the "father of Silicon Valley" because he encouraged his Stanford engineering students to start companies that would create jobs for graduates. Two students who did, Bill Hewlett and David Packard, founded HP in 1939.)

The publisher turned to electrical engineering Professor R. J. W. Koopman of Washington University in St. Louis about preparing a textbook on the subject. Shortly thereafter, however, Koopman succeeded Roy Glasgow as department chairman, and progress on a manuscript slowed dramatically. Koopman suggested to McGraw-Hill that they ask a former Washington University instructor, Walter R. Evans, whether he would be interested in writing a book using course notes developed at Washington University, Emerson Electric, and North American Aviation.

On August 2, 1949, McGraw-Hill editor John Wight wrote to Walt:

> *We have nothing in the way of an immediate prospect which could be considered competitive with the distinguished book by Brown and Campbell. I gather your work ...would include material both below and above the academic pitch of Brown and Campbell. ...I know that Dean Terman will be most pleased to have an opportunity to look your material over.*

On August 21, 1949, Walt sent sample material to Wight, with a warning that his approach would be unconventional:

> *The book's main purpose is to demonstrate the root locus method. The key idea is a simple graphical plot to locate the roots of the characteristic equation ... If the book could be made as competitive as the method itself, ... [we] would be both pleased. Frankly, however, there is a gamble involved for each of us in that books emphasizing the physical picture of a subject are in the minority relative to those for which mathematics predominates. I am personally convinced from teaching General Electric's Advanced Engineering Courses and undergraduate courses at Washington University that the students themselves want the physical picture.*

Both the author and the publisher sought an early completion date. Editor John Wight expressed McGraw-Hill's business goals in an October 1949 letter:

> *The Servomechanisms market is growing with great rapidity. There are, as you know, good books already on the market and certainly more will be written. For this reason, we are most anxious to get your book underway and hope you will find the time to finish the manuscript at a reasonably early date.*

On November 1, 1949, Walt signed a contract, already signed by McGraw-Hill President Curtis Benjamin, for a book to be titled *Control System Dynamics*.

The First Draft (January 1950–March 1951)

Walt turned thirty on January 15, 1950. He marked the occasion by writing to Wight, suggesting a meeting while in New York for the AIEE winter meeting. There, he was set to present his groundbreaking paper, "Control System Synthesis by Root Locus Method." During their meeting, he estimated that a first draft could be completed by June 1.

Fully committed to the project, Walt set up an office in a neighbor's spare bedroom. Each morning, after pretending to leave with his carpool, he simply walked across Maple Street to his writing space. By mid-June, he had completed a first draft and began revising it with input from colleagues at NAA.

In December 1950, he submitted sample chapters to McGraw-Hill and committed to a full manuscript deadline of March 1, 1951. Initially estimating a 170-page book, he soon realized that with 488 figures, the total length might reach 300 to 350 pages. Concerned about affordability, he hoped the book could be priced at around $5.00.

McGraw-Hill Editor Ken Zeigler, who had replaced Wight, responded within a week: "We're very pleased with your progress and hope our Editing Department can handle it as is."

By March 1951, Walt knew of three prospective servomechanism books: one by Harold Chestnut, one by Floyd Nixon, and one by George Thaler. Recall that Harold Chestnut had been Walt's supervisor at GE in 1942. Nixon and Thaler sent letters to Walt, informing him of their intentions to have chapters on the root-locus method in their books. Walt had copies of AL-787 sent to them. Nixon and Evans exchanged letters sharing their mutual difficulty of choosing a title. Walt quipped, "most fellows I know just name the authors and probably couldn't name the title on a bet." Thaler's letter included a request: "If it is possible, would you please send me a copy of your notes so I can check my work and extend it."

Walt, perhaps seeking a differentiator that would provide *Control-System Dynamics* a competitive advantage, proposed bundling the Spirule with his book. McGraw-Hill rejected the idea, citing cost concerns. Zeigler explained, "It would raise expenses (and thus the book's price) without necessarily increasing sales."

Moreover, time to market was critical. Zeigler told Walt about upcoming books, including works by Ahrend and Taplin, by the aforementioned Chestnut and Mayer, and Thaler and Brown—with the first expected within six months. With competitors crowding in, Walt had to complete *Control-System Dynamics* quickly if it was to establish itself as a leading text on servomechanism design.

Spring 1951

By the spring of 1951, Walt had completed the remaining sections of his manuscript. He also adjusted notation and equations to align with McGraw-Hill's editorial standards. However, he disagreed with Editor Ken Zeigler's stance against including the Spirule. "Most engineers here at North American Aviation believe it should be enclosed," Walt wrote. "Some have even suggested that readers might resent having to make a separate purchase."

Two years after signing their Memorandum of Agreement, McGraw-Hill informed Walt that their technical reviewer recommended significant revisions, including the addition of problem sets. Interestingly, the reviewer supported including the Spirule, writing, "The instrument called the 'Spirule' should probably be included with the text since its use is treated rather fully …"

The Laplace Transform Controversy

Beyond the absence of any problem sets and the problematic Spirule question, the reviewer also questioned Walt's decision to introduce transfer functions without the Laplace transform. But Walt stood firm. He had deliberately avoided the standard approach, believing that students often memorized rules rather than developing a deep understanding of system behavior.

Consulting editor Frederick Terman tactfully sided with the reviewer, "The suggestion that the Laplace transform might be used is one that the author should consider seriously."

In the following week, Walt prepared a defense of his case in an essay he called *An Opinion Concerning the Laplace Transform*. It is the most passionate discussion Walt ever wrote about the education process:

> *The Laplace Transform is admittedly the simplest way to <u>present</u> many conclusions regarding transform functions; I do not believe however that it is the simplest way for the student to <u>understand</u> them. Of course many students do not seek understanding; show them a routine, assign a few problems for practice, give them a straightforward exam, and if they pass it, they leave the course feeling educated.*
>
> *If however some of the problems are not essentially duplicates of the homework problems, they fail miserably and then complain of the*

> *examination as being "unfair." The fault does not lie entirely with the students and not with the Laplace Transform itself, but I do not believe that the above method of education should be encouraged. The graduates of such a system are of little more value to industry with the course than without it. If he has to solve a problem, he must take time to reestablish the routine. If he makes an algebraic error, he may arrive at a ridiculous result without recognizing it. ...*
>
> *Admittedly there is no foolproof way of presenting a method. I have already had the disgusting experience of having an engineer come in and ask advice on some detailed phase of plotting a root-locus plot only to find out gradually that the method had nothing to do with the problem he was supposed to solve!*
>
> *Fortunately however there are many students who make very reliable engineers once they understand the fundamentals of their subjects. I have tried in Chapter IV to present a chain of reasoning which will establish transfer functions and all their properties on the sure foundation of the solution to a few simple differential equations by the classical method of trial and error. This chain of reasoning led to the development of Root Locus and it has often been of value in explaining the method. If the method of presentation is poor, I would surely appreciate criticism; but to substitute the Laplace transform in my present state of understanding of it would be unfair to the conscientious student for whom the book is intended.*

Concerned about the sharp tone of his argument, Walt wired Zeigler, instructing him to destroy the essay before delivering it to Terman. However, he kept a copy for himself. Two weeks later, he sent Terman a softened version of his position, citing examples of students' struggles with the Laplace Transform and how it hindered their problem-solving abilities.

The Spirule: Final Decision (December 1951)

As Thanksgiving 1951 approached, two years after Walt signed the Memorandum of Agreement, the fate of *Control-System Dynamics* remained uncertain. Would McGraw-Hill continue with the project, or would they favor the competing servomechanism manuscript by Thaler and

Brown? The publisher's technical reviewer had criticized Walt's presentation as unconventional, and doubts lingered about including the Spirule.

But just two days before Thanksgiving, Ken Zeigler offered some encouragement:

> *We have received praise of your ability from several sources recently. The comments we have heard support the confidence we have placed in your successful completion of the project.*

Ten days later, Zeigler relayed Dean Terman's assessment:

> *Evans' viewpoint on the Laplace transform is a perfectly reasonable one. I think that some explanation about the place and usefulness of the Laplace transform and alternative methods should be incorporated somewhere in the book. If this is done, and Evans then wishes to leave out the Laplace transform completely, I would say fine and proceed on that basis.*

Terman's endorsement reinvigorated both Walt and McGraw-Hill, providing the momentum needed to move forward.

The AIEE Annual Conference was scheduled for December 6–7 in Atlantic City. Walt proposed an in-person meeting in New York on December 5. It would be his first opportunity to meet Zeigler and other members of the editorial team. In his letter, he once again raised the question of including the Spirule:

> *The matter of enclosing the Spirule in the back cover of the book should perhaps be decided now, or at least all the facts presented. I will bring several of them with me also.*

In New York, Walt met the two men most involved with his book—Ken Zeigler and Jeff Norton, whom Zeigler had assigned as editor. Walt's handwritten notes from their meeting suggest that all parties anticipated an early 1953 publication, with an initial print run of 2,000 to 3,000 copies. However, after considering that including the Spirule would raise the book's price by $1.50, Zeigler once again advised against it.

Walt's notes reflect the final decision: "Recommends I sell it separately (may make as much as from royalties)."

With that, the Spirule matter was settled.

Perhaps the clearest explanation of Walt's motivation for writing the book is found in a January 1952 letter to Professor Elias Sabbagh of Purdue University. Walt had sent him a draft copy of the textbook upon learning from NAA recruiter Ray Hamada that Sabbagh was "building up a strong servomechanism program." In the letter to Professor Sabbagh, Walt shared his own experience:

The first observation in seeking to teach the root locus idea to engineers at North American was that far too many recent graduates had learned a routine for use of the Laplace transform or for making logarithmic plots but had little understanding of what they were doing. ... I decided to go all the way back to the basic fundamentals including even the basis for the choice of "e." The subject is then developed until the final problem is that of manipulating determinants in order to study the effect of various coupling effects in the roll-yaw motion of an airplane. Frankly the root locus idea is primarily just an excuse for writing a book in which physical concepts are dominant with mathematics secondary. There are also many criticisms that the book has too many "novel explanations," but I have long ago learned that you can't please everybody.

Revised Priorities and a Missed Deadline

The prospects for *Control-System Dynamics* seemed bright as the winter of 1952 transitioned into spring. Walt had completed all the revisions requested by Dean Fred Terman. To his pleasant surprise, the first Spirule production run of 513 quickly sold out, defying his expectation that strong book sales would be the key to strong Spirule sales.

When Arline and Walter celebrated their tenth wedding anniversary on April 11, they had another cause for celebration—Arline was about to enter her third trimester with a mid-July due date for their third child. However, when Arline's blood pressure spiked in May and she was hospitalized, her health became Walter's only concern. When a scathing review of his manuscript arrived from McGraw-Hill publishers on June 5, he set it aside.

At Murphy Memorial Hospital in Whittier, doctors, concerned for her health, decided to deliver the baby weeks before its July due date. On June 12, Arline delivered a 5 lb. 4 oz baby girl, Nancy Arline Evans. To

Arline and Walter's great relief, the premature baby proved to be healthy. Arline recovered rapidly.

Nancy Arline Evans arrived on June 12, 1952. Pictured with Randy and Greg.

Response to the Scathing Review

As for the scathing review that had arrived a week before Nancy's birth, McGraw-Hill's Ken Zeigler simply wrote, "I am afraid that you will find his comments rather disappointing. It may be that he is a particularly severe critic."

The reviewer began by expressing his high expectations after meeting Walt:

His personality and his ability to think quickly have made a favorable impression at professional meetings. He has recently been appointed a member of the AIEE Committee on Feedback Control Systems. Because of this personal background, I looked forward to seeing a text by Mr. Evans.

Following those positive comments, he wrote:

The basic conclusion of my review is that this book, if published in present form, would not do justice to the author's reputation. ... It is not an easy task for a technical industrialist to find time to write a book. It is also difficult for someone as steeped as I in the field to review such a book with the proper perspective. While I think a stranger to the control field with only a general technical background would have a very difficult time with Mr. Evans' presentation, I am not at all sure of this. I am quite sure, however, that a scholar in the field would find too little new in this book by Mr. Evans.

Walt, who had surely reassessed his priorities after Arline's health scare, dismissed the criticisms in a June 24 response to Ken Zeigler:

The [critic's] comments … are not surprising; the disappointment to me is the time consumed in receiving reviews by one critic after another. The style and presentation have been set by five years of study and teaching in the General Electric Advanced Course, in which the prime object was understanding rather than rigor or elegance.

McGraw-Hill had made clear that, to secure university adoptions, the book needed to be available by the spring of 1953, when professors would decide on textbooks for the fall semester. Zeigler, after consulting Dean Terman, instructed Walt to address only the reviewer's most essential concerns to avoid further delays.

But Walt's priorities had changed. Three months came and went. *Control-System Dynamics* missed the deadline for fall 1954 course selections. Perhaps Walt felt he had short-changed his responsibilities as husband and father too many times in the last five years revising manuscripts to satisfy reviewers and critics at the AIEE and McGraw-Hill. Shortly after he submitted his revision, it came back in November with instructions to retype portions of it.

Walt may have missed its deadline, but other authors had met theirs, including George Thaler at McGraw-Hill. In 1953, McGraw-Hill's *Electrical and Electronic Engineering Series* published his book as its first servomechanism textbook. Its "Chapter 14: The Root Locus Method" was based in part on the AL-787 report Walt had sent Thaler in March 1951. However, the book never once mentions Evans by name.[12]* The omission of "Evans" was never an issue for Walt. He read Chapter 14 and sent George Thaler improvement suggestions for future book printings.

Thaler became one of Walt's most prominent supporters. After Walt's

---

* Except for references to Walt's AIEE papers buried in a 206-entry bibliography in Appendix F.

1987 Rufus Oldenburger Medal ceremony, Arline wrote to Thaler: "I wish to thank you for all your efforts on Walter's behalf."

Pushing Toward Completion (1953–1954)

On February 6, Walt sent the retyped manuscript via Railway Express. His cover letter ended with a familiar refrain: "I hope that you will now find the manuscript completely satisfactory."

But instead of final approval, McGraw-Hill's new copy editor, Jeff Norton, redirected Walt's attention to minutiae—particularly the rules for hyphenation. Norton insisted on inserting a hyphen between *Control* and *System* in the book's title and between *root* and *locus* when referring to the *root-locus method*. With no fight left in him, Walt readily acquiesced.

Unsatisfied with some of Walt's sentence structures, Norton also suggested that he share in the $200 cost of the copy editing. More delays followed. The copy editor fell ill for an entire month, stalling progress further. Meanwhile, Walt faced a deadline for a new root locus paper targeting problems of interest to mechanical engineers. The urgency that had once driven both Walt and McGraw-Hill had long since dissipated. As 1953 ended—four years after the project's inception—*Control-System Dynamics* remained unfinished.

On February 19, 1954, the manuscript was finally sent to the printer. Galley proofs arrived in mid-April, and by September 30, the first printing of 2,500 copies was distributed. McGraw-Hill advertised the book extensively, with promotional materials appearing in technical journals and trade publications.

Roots of Root Locus Remembered with Gratitude

In the preface of *Control-System Dynamics*, Walt expressed gratitude for the education he had received:

> *The author is grateful to Profs. Roy S. Glasgow and Frank W. Bubb, both formerly of Washington University in St. Louis, for their emphasis on understanding rather than superficial knowledge. The General Electric Advanced Engineering Program continued this emphasis, furnished extensive practice in the solution of problems, and provided association with such fellow students as G. E. Walter.* [29]

He also credited Paul Profos of Switzerland for his initial inspiration, writing: "Specific references are given, [such as] the basic paper by Paul Profos on conformal mapping to find roots."[29]

On November 28, 1954, Kenneth Zeigler wrote to Walt requesting corrections for a second printing. Walt responded on December 11: "The news that the total requirement for *Control-System Dynamics* might exceed 2,500 by the time second-semester orders are placed is the biggest surprise that I have had for years. ... Please have 12 (books) sent (to me) as quickly as possible because I want to send some of them out at Christmas."

Walt had mixed feelings about his publishing experience. He was satisfied that the book was heading for a second printing. Yet, in one of the books he sent as Christmas gifts, he wrote this note to his brother and sister-in-law: "Dear Sam and Betty, If Mrs. Levy, my English teacher at Soldan High had had to approve this book—it might never have been published." He then added a more somber comment. "It's too bad when you reach a goal like this, it doesn't seem worth reaching as much as when you started. Your lil brother, Walt."

Despite a promising start, *Control-System Dynamics* soon faced stiff competition. By 1955, multiple textbooks had adopted and expanded upon the root-locus method, limiting the book's impact. Sales declined, and between 1954 and 1957, McGraw-Hill sold approximately five thousand copies. McGraw-Hill found that most sales came from outside the academic market:

> *Comments I have received on the book are: "too brief" and "excellent after you already understand the subject!" "I was amazed in teaching the course again how much students want to have every step detailed for them. The unfortunate thing is they won't get problems that clear cut in industry and had better get into the habit of filling in the details.*

Walt remained proud of his book, which emphasized practical understanding and built upon principles learned from his mentors. In February 1955, his Washington University professor, Frank Bubb, now Chief Scientist, Office of Air Research, Wright-Patterson Air Force Base, having received one of the copies Walt had sent as gifts, wrote both to thank Walt and to express his support for the pedagogical approach Walt had taken in the book:

> *It seems to me that anyone who undertakes an exposition of systems analysis and synthesis should not only explain the mathematical technique but also show how to attain true understanding. I have debated this issue with Stark Draper, Gordon Brown, Bill Ahrendt, and others. Most of them argue that understanding comes through some mysterious process of osmosis alongside diligent application of design techniques. This is, of course, pure rot.*
>
> *A sufficient number of parallel discussions—one in the time domain, the other in the transform domain—can bridge this gap. ... Your book's instructive parallel discussions are commendable ... I am sorry that none of the journals asked me to review your book. I really would have spread it on.*

Walt expected textbook sales to drive the acceptance of the root-locus method. It did not, nor need it have. By 1954, the root-locus method had already secured its place in history.

Three events that preceded publication of *Control-System Dynamics* had made its disappointing sales irrelevant to the acceptance of his method.

1. AIEE Paper: "Control-System Synthesis by Root Locus Method"

The March 1950 publication of Walt's paper in *AIEE Transactions* precipitated a virtual avalanche of papers—some important extensions, some merely derivative. By one count, Walt's paper had been cited in sixty-one others by 1957. In 1951, the Institute of Aeronautical Sciences scheduled an entire session at its annual conference in Los Angeles to discuss the root-locus method.

2. Fourteenth Wright Brothers Lecture

The impact of the September 1951 publication of Dr. William E. Bollay's Fourteenth Wright Brothers Lecture, originally given on December 16, 1950, at the Institute of Aeronautical Sciences in Washington, D.C., cannot be overstated. Bollay, founder and head of the Aerophysics Laboratory at North American Aviation, was a nationally known and respected figure. He was in a position to influence perceptions about root locus across the entire aerospace community. In preparation, Bollay stayed late at NAA and met with Bill Mullins to understand the root-locus method thoroughly.

Dr. Bollay began his remarks on his topic, "Aerodynamic Stability and Automatic Control," with words chosen to get the attention of the audience:

> *The practical achievement of satisfactory stability and control is probably the greatest contribution of the Wright Brothers in the development of the airplane. ... At present, the airplane is going through another period of transition similar to that a half-century ago. ... Probably the major development in aircraft stability in the past 10 years has been the evolution of analytical and experimental techniques that permit an engineering calculation of the motion of an airplane when under the control of an autopilot. ... The principal techniques that have made this development possible are the following: ... The graphical methods of analyzing the dynamic performance of a system, particularly the diagrams associated with the names Nyquist, Bode, and Evans.*

There it is! The equal billing Walt received referred to a March 1952 letter to friend Jack Clark, "[Bollay] gave root-locus billing in parallel with the classical stuff, and that really got it started."

Successors of the Wright Brothers

Uncommented upon by Walt, but more significant to the flying public, was the "billing in parallel" Bollay gave these three little-known engineers with the legendary Wright brothers. Viewed in their entirety, Dr. Bollay's opening remarks on December 16, 1950, may be summarized as follows: Harry Nyquist, Henrik Bode, and Walter Evans deserve recognition for ensuring the stable flight of high-speed aircraft, just as Orville and Wilbur Wright deserve the recognition afforded them for ensuring stable flight on low-speed aircraft.

But Bollay did not stop there. He then compared the relative utility of Walt's root-locus method to the analysis methods of his two servo-analysis "siblings":

> *The Evans root-locus method presents directly a complete picture of the stability and transient response characteristics that are most important ... The root-locus gives the roots of the closed-loop system directly and by a simple calculation, the transient response. The degree of stability can be read from the root-locus directly. Complicated systems can be set up in such a fashion that the effect on transient response and*

stability of changing any parameter can easily be visualized. There is no ambiguity in the interpretation of plots even for complex systems having any number of roots and poles in the right or left half plane. [26]

When William Bollay's December 1950 Wright Brothers Lecture appeared in print in the September 1951 *Journal of Aeronautical Sciences*, it did more than put the root-locus method on the map. It put it on a veritable mountain top!

3. A Chapter on Root Locus in Every New Control Theory Textbook

In 1953, McGraw-Hill published *Servomechanism Analysis* by Thaler and Brown. Two years later, Wiley came out with *Automatic Feedback Control System Synthesis* by John Truxal. Prentice Hall, Van Nostrand, and the aforementioned published still more textbooks in the 1950s and 1960s, all with a chapter on root locus. Many college bookstores soon followed the example set by UCLA and UC Berkeley, where Joe Beggs and Otto Smith required use of Spirules in classes.

And so, as it turned out, sharing his servomechanism course notes in 1951 with George Thaler, John Truxal, and Floyd Nixon was not only generous, but it was also smart business. In Frank Bubb's words, these three men "spread it on." To cite one example, John Truxal's textbook stated:

> *Among the many methods presented, the root-locus approach stands out, for it combines the theoretical advantage of simultaneous control over both transient and frequency responses of the system with a strong appeal to the designer, an appeal that derives from the simplicity of the method as well as the logic of the underlying the approach.*[35]

An Inflection Point

Recall Frank Bubb's wish that he had had the opportunity to "spread on" *Control-System Dynamics*. Walt believed that a relatively small number of engineers, especially users, "spread on" the root-locus method. When the Spirule Company began receiving orders, Walt sent personal thank you notes. In such a thank you note to Louis Wadel, Walt shared his observation of auto-navigator designer Ken Jackson, adding a characteristic wisecrack at its end: "… interest tends to arise because of an enthusiast like Ken Jackson working out a few examples. …Thanks for the [Spirule] order. If this keeps up I'll break even!"

There it is: a flash of humor! A constant for Walt was his sense of humor. It attracted Arline Pillisch at Soldan High School and it was the quality most often mentioned by NAA colleagues. Its absence in Chapters 8 and 9 is attributable in part to the frustrations Walt experienced fending off critics and in part to our extensive use of contemporary correspondence in the service of historical accuracy. However, *Into Stability* has now reached an inflection point.

Parts I and Part II have held a tight field of view, capturing the first half of Walt's life in crisp historical detail through the sharp focus of quotations from historical documents. With the publication of *Control-System Dynamics*, the root-locus method became less dominant in Walt's life.

The narrative needs a wider angle lens, a more open aperture, and more relaxed exposure, softening and expanding the image. In that broader frame stands Arline—the steady presence who managed the household, raised their children, oversaw the Spirule Company's daily operations for two decades, and, in the last nineteen years of Walter's life, became his full-time caregiver. Her constancy was the quiet foundation of his stability.

Moreover, the narrative will be composed in a different light, drawn from what Lincoln called "the mystic chords of memory." It will gather the recollections of colleagues in Chapter 10, delineate Arline's roles in Chapter 11 and the Epilogue, and draw upon the author's memories in the Epilogue and "My Stories of Dad." Together, taken from different vantage points and perspectives, they will form a fuller, more expansive portrait of Walter R. Evans, not only as an engineer, but as a man whose life's balance rested on the enduring partnership that sustained him.

Cardinals by Walter R. Evans, February 20, 1987, Portfolio #155, Vol. 2

Part III
STABILITY

*Now, here I cannot help
but say a few words about Arline.
My dear, for me, you are truly wonder woman.
What role models you and Walt are for us.
You lived your marriage vows day by day,
for better or worse, in sickness and in health.*

Dr. Edward H. Bloomfield
July 1999 Memorial Service for Walter Evans

Chapter 10

Autonetics (1955–1959)

A Transition Year for Walt (1954)

After his decade of accomplishment (1944–1954), Walt had firmly secured his place in the history of control theory. The publication of his textbook coincided with those of many other books of the time, all of which dedicated extensive sections to the root-locus method of design.

On the home front, two significant events made 1954 a milestone year in his life. First, in March, Arline gave birth to their fourth and last child, a son, Gary.

Four Evans Children in 1955.
Randy (11) Gary (1), Nancy (3), Greg (8)

Then, in August, the family moved from their modest two-bedroom home on a tenth of an acre on Maple Street in uptown Whittier, to a custom-designed five-bedroom home on a half-acre in East Whittier. Walt and Arline personally designed the floor plan. Walt even built a scale model of the house himself. The contractor, Lew Bonser, was known for his high-quality construction. The flooring was made of 2 x 6 tongue-and-groove lumber. Walt bought a 1955 Buick station wagon. It would serve as the family car for many road trips.

In 1954, the *Brown v. Board of Education* Supreme Court decision began dismantling racial segregation in schools, signaling the rise of the civil rights movement.

Popular culture also saw significant change in 1954. Elvis Presley had his first hit single with "That's All Right." Television shaped American households with wholesome family shows like *Father Knows Best* and *Lassie*, while *The Adventures of Ozzie and Harriet* portrayed idealized suburban life. Color TV technology was advancing, setting the stage for new entertainment experiences, and Walt Disney broke ground for Disneyland, symbolizing postwar optimism. This combination of social shifts and cultural milestones marked 1954 as a transformative year in modern history.

1954 marks a personal transition as well. I turned seven in 1954. Up to this point, I have no firsthand memories of the events described in this book. That changed in 1954. Thus, it is now appropriate for me to continue this narrative in first person, rather than third.

Aerophysics Lab Company Parties

Miriam and John Moore, Mom, Jesse and Dorothy Bowman, Dad, and Fred and Mona Eyestone were pictured at a company-sponsored dinner. Dad could be quite animated at such events.

One of Dad's colleagues was George Anderson. He and his wife hosted an annual Christmas party at their Whittier home. One year, a spouse of one of Dad's colleagues took Mom aside as the evening wore on and said to her, "Don't you think Walt has had enough to drink, dear?" Neither he nor Mom drank, and he had had not one drop of alcohol. He was exuberant.

AUTONETICS (1955-1959)

John R. Moore, Marketer Extraordinaire

In the late 1950s, whenever I asked Dad what he did at work, I got the same response: "I am one of a hundred engineers trying to keep Johnny Moore an honest man." In those heady days of the 1950s, beyond being a consummate engineer and recruiter, Johnny Moore was also the best marketing executive at North American Aviation. His frequent trips to Washington, supported by outstanding technical work in California, contributed to the growth of the Aerophysics Lab. Moore counted on the ability of his stable of engineering talent to deliver on his promises—especially those associated with the Navaho Missile program and the Polaris submarine fleet.

Over just ten years, the program accomplished several technological firsts.[31] The X-10 test drone became the first turbojet-powered vehicle to exceed Mach 2 and the first aircraft to fly a complete mission under inertial (computer) guidance. Later, the G-26 ramjet-powered vehicle became the first jet aircraft to reach Mach 3 and an altitude of 77,000 feet.

The accomplishments of the Aerophysics Lab laid the groundwork for dramatic growth in inertial guidance technology. This technology evolved through multiple generations, from XN-1 to XN-6. As described in Chapter 6, the XN-1 was the first inertial guidance system to be tested in an airplane.

> *The guidance and control computer for the Navaho was completely digital, with a rotating disc, non-volatile memory. It made its first flight in 1955 as the first transistorized digital computer with a rotating disc memory ever flown.*[31]

The Navaho program is one of the least-known yet most important early missile programs in the United States. The project achieved advancements in every discipline of engineering and electronics. It introduced the airborne digital computer, modular electronic circuitry, and the first fully inertial navigation system. James Gibson wrote an excellent reference: *The Navaho Missile Project*.

In 1955, North American Aviation restructured its MACE (Missile and Control Equipment) division, which focused on guided missile systems and control equipment. This restructuring led to the creation of Autonetics, a division specializing in avionics, guidance, and control systems. As general manager of Autonetics, John Moore then led a division of more than seven thousand people.

Nautilus 90 North: August 3, 1958

Autonetics continued to expand, securing ambitious contracts for equipping the navy's Polaris submarine fleet and the air force's silo-based Minuteman missile installations. In 1958, the XN-6 navigator, originally developed for the Navaho, would guide the navy's nuclear submarine Nautilus to the North Pole. When I interviewed John Moore in 2003, he expressed how he was proud that he and the company had had the foresight to ensure the XN-6 inertial navigation system would fit through a submarine hatch. In Moore's memoir *Legacies Left by Former North American Aviation Experts*, he wrote:

Tom Curtis onboard the Nautilus en route to North Pole guided by the XN-6 inertial guidance system

> *In the early 1950s, our gyro engineers came up with a paired, reversing gyro design which automatically canceled almost all of the error torques dependent upon the gyro angular momentum vector. This six-gyro stable platform, in which three gyros performed the h function while the other three were reversing their angular momentum vectors, provided a concept which seemed to meet the Navaho accuracy requirements without depending upon a star tracker called 'NAVAN' [North American Vehicle Auto Navigator]. Tom Curtis was the project engineer on the NAVAN, which was selected as the guidance system for the Navaho missile. This concept had so much promise that we recognized its potential application as an inertial navigator for submarines. Accordingly, we established as one design criterion that the stabilized platform and associated guidance electronics had to be small enough to pass through a submarine hatch.*[19]

Autonetics Growth Came with a Price

The Aerophysics Lab's entrepreneurial culture, which encouraged individual accomplishments, gradually gave way to a more bureaucratic culture.

John Moore admitted that mistakes were made, some of them his. He wrote:

Autonetics had gone from a nadir of 5,400 people at the beginning of 1958 to a requirement for rapid growth. ... It soon became evident that for such a buildup as all of these new programs required, the original Autonetics organization structure was totally inadequate. Accordingly, in 1959, Autonetics was reorganized along product lines.

However, I made the mistake of affecting this organization without first changing the existing organization's operating policies, procedures, specified authorities, and responsibilities to relate to the new organization. This produced so much chaos that the division escaped meltdown only because of the dedication, skills, and stamina of its key people ...[19]

One consequence was a loss of talent. Some engineers went to other companies, and others struck out on their own, most notably Henry Singleton, who left and founded Teledyne, becoming a multi-millionaire.

The rapid growth of Autonetics was a mixed blessing for Dad. In college he had followed Roy Glasgow's advice and switched out of the engineering administration track taken by his father. Yet, his father's title "Vice President" may, in the back of Dad's mind, have been a standard against which he compared himself. It is a force field for a son. I have felt its power. Dad escaped that force field in 1959. At the age of thirty-nine, halfway through his career, he decided to look for a work environment in which he would be allowed to do technical work full-time. Full stop.

There were too many colleagues that admired him as a manager for him to achieve that goal at North American. He would have to look elsewhere.

He reached out to Robert Cannon, now a professor at Stanford, to advise him on options in the Bay Area. Cannon responded with four: Lawrence Livermore Labs, Stanford Research Institute (SRI), Lockheed Missiles and Space Division in Sunnyvale, and Stanford. Dad and Mom flew to Palo Alto on July 22.

Meanwhile, Dad explored his options near Whittier. Ultimately, Dad chose the company he believed would afford him interesting technical problems, have a first-class engineering staff, and be within commuting distance. He chose Ford Aeronutronic in Newport Beach, California—a forty-five-minute commute.

INTO STABILITY

The rest of this chapter has a variety of materials from this time of transition. Insight into Dad's views on his day-to-day job is conveyed in a 1957 letter to his boss, Fred Eyestone. A different take is found in one he wrote to a colleague on his last day at Autonetics: August 7, 1959. I share the memories his colleagues shared with Dad or me. (Still more remembrances are in Appendix 8.)

I'll then conclude Dad's North American Aviation chapter by revisiting his relationship with the person who, more than any other individual, launched his career: John Moore. At the end of the chapter is Dad's letter supporting Moore's nomination for an honorary degree from Washington University. In 1988 and 1990, John R. Moore and Walter R. Evans received Engineering Alumni Achievement Awards from the McKelvey School of Engineering at Washington University in St. Louis.

Dad's Letter to His Boss, Fred Eyestone

In September of 1957, Dad received a bonus check from his boss, Fred Eyestone. Afterward, he sent Fred a three-page, typewritten letter. It contained a confession, a tactfully worded pushback to management's goals, and several constructive suggestions. Dad liked Fred. Outside of work, Dad and Mom played bridge with Fred and his wife Mona. They all were residents of Friendly Hills in Whittier. A reading of this letter is as good a way as I know to understand how Dad approached his work. In order to be more readable, I used AI to condense it. But even with the rewrite, I hear Dad's distinctive voice.

Dear Fred,   *Sept. 30, 1957*

Receiving today's bonus check reminded me how much our work has contributed to the company's success. It also prompted me to reflect on the challenges we've faced—especially in reducing costs while maintaining high engineering standards. Discussions at work are often too rushed to go into detail, so I'm taking this opportunity to outline my thoughts.

One of my biggest regrets is not identifying a major issue with gyro drift under vibration sooner. The first clue came from Romberg's tests on the X-4, but I initially misdiagnosed the problem. It wasn't until Tom's tests confirmed similar results that we realized the issue lay in servo frequencies. Thanks to Dave Chandler's low-noise circuitry, we finally solved it. However, this should have been caught years earlier, which would have saved significant costs in testing and unnecessary hardware.

We've faced increasing pressure to cut costs by standardizing hardware, treating components as off-the-shelf items. At one meeting, management seemed to favor this approach, which was discouraging given our hardware's state. Fortunately, after thorough discussion, our proposed changes were approved. This reinforced my belief that we must stick to sound engineering principles, even when they are unpopular.

There have been ongoing issues with angular readouts across shock mounts. Ted investigated and determined that inconsistencies in mount stiffness could lead to errors of up to 20 arc seconds. This highlights the need for thorough analysis before adopting new designs, rather than blindly following external recommendations.

Temperature control and friction gearing are also areas where we've made progress. Improving these systems can yield greater cost savings than simply trying to streamline shop processes. There's been criticism that our group doesn't focus enough on producibility, but the real issue is balancing design integrity with manufacturing constraints. Some policies—like component detail specifications—were meant to improve planning but have become bureaucratic obstacles instead. We need better coordination between design and production. My suggestion is to assign Wienecke to bridge the gap, allowing real-time feedback rather than waiting for problems to surface after drawings are finalized. This would be especially useful now that the main shop is in Compton.

I believe we often see different sides of the same problem, which is why I wanted to clarify my perspective. Our work has proven that sound engineering, rather than shortcuts, leads to real cost savings and technical success. I hope this letter helps move us toward better solutions.

Sincerely, Walt Evans

Dad's Last Week at Autonetics: Final Thoughts in His Own Words

In his last week at Autonetics, Dad wrote several letters, including one to John Moore, which, regrettably, is missing. We have a short note to Bob Moore, a colleague and neighbor. It reveals what was on his mind when he resigned.

To: R. E. Moore   August 1959

The attached memo reminds me of some thoughts which might be well expressed at this time. Several people, on returning from trips, have been surprised at how frequently my name is known at universities. It is due almost completely to the popularity of servo as a subject and the novelty of root locus as a method.

I am always leery of anyone who is too enthusiastic about the method because it suggests that they are not aware of the whole problem. They might also be subject to being "fad happy" without really thinking about the problems themselves and the many possible solutions. Some clues on the adoption of root locus can be gained from the fact that Spirule sales have gradually climbed to the present figure of about five thousand Spirules per year.

My present problem is that bookstores receive them at $2.10 and mark them up, typically to $3.50, and in one case to $3.95. Arline and I use the approach of notifying the bookstore of our direct sale figure of $3 and likewise notifying the professor involved. We guarantee the bookstore the return privilege in case they are caught with an overstock as a result.

Profit to myself runs about $1,000 to $2,000 and my main interest is to provide a useful gadget rather than make a killing. The prospect of students being forced to buy something they don't want at prices they figure are too high worries me. So, if you pick up any information along this line, I would be glad to hear it.

Walt Evans

North American Aviation Colleagues' Memories of Walt

Walt's friends and colleagues shared memories of Dad on four occasions:

AUTONETICS (1955-1959)

1. 1985: Dad's official retirement from Autonetics
2. 1987: Rufus Oldenburger Medal Award Ceremony
3. 1999: Dad's passing on July 10
4. 2003: My requests for remembrances of Dad and accounts of events

Extended recollections by Gordon Walter and Jeff Schmidt are found in Appendices 3 and 5. Brief anecdotes like the following are in Appendix 8.

Hal Engebretson:

"I started work for North American Aviation in 1954. My supervisor was Sam Carlson and the Group Leader was Walt Evans. There is certainly no question that Walt was an outstanding technical talent. Many people respect Walt for his outstanding technical ability. I also respect him as an outstanding manager of people. When I started to work, it became obvious that if you worked in Walt's group, he supported you. The message was that one did not have to worry about trying to do a good job—if you did your best, Walt would support you. That attitude resulted in some very good people developing to do very good work."

Jeff Schmidt:

"Walt was a walking engineer's handbook. You could ask him anything from heat transfer to hydraulics. He would pull a few fundamental constants from memory, apply a few conversion factors, do some math in his head while telling you what he was doing, and come up with an answer. In summary, Walt was a very creative person with a great sense of humor."

Sam Carlson:

"I had been interviewed at Purdue by Ray Hamada, who was on a recruiting trip for North American. I told Ray that if I could work for Walt Evans, I would accept his offer. I consider myself most fortunate to have had the opportunity to get started in my career in the aerospace industry working for Walt. All of us who worked with him learned so much: pride in technical excellence, respect for absolute integrity, and satisfaction in accomplishing extremely challenging goals through dedicated individual and team efforts."

DeWitt Lyon:

"Walt was always so careful to remember who did what and to encourage and give credit where credit was due. The combination of humility and intellect was exemplary and, of course, always coupled with good humor. One example of his outstanding memory was not only the ability to remember humorous situations and jokes, but also to remember to whom he had told them."

Robert Nease:

"Walt was truly one of a kind—not sure they make guys like him anymore. Although I never worked directly for Walt, he was one of the first of the 'old timers' that I met in Downey (Mitz was a friend from school and instrumental in my coming to work for Autonetics). We had lots of good times in Downey, a few years together in Newport Beach, and then quite a while in Anaheim. We shared an interest in (very amateurish) athletics—seemed to have something going most of the time, with lots of that while we were at Ford. But the big hero in this story is Arline. You meant so much to him and have spent most of your life taking care of him/family, without a lot of fanfare and praise. Walt was so proud (and rightly so) of his family and always told me that you were responsible for anything good which happened."

Robert Cannon:

"At NAA, carpools could park closer to the entrance. I came up behind Walt and beeped my horn. Walt did not turn around or step to the side; he jumped onto the back of whoever was walking in front of him. Walt always loved a technical argument ... and he loved to win. I can still hear him as he points to a genius pad totally covered with scratches, saying, 'There's no chance for any error here!' For the center of discussion in our office would be a huge pad of blank paper called a genius pad in the center of the conference table, which was between our three desks, and there was also a fourth chair in the doorway for guests.

One day, Walt and a guest were going at it hot and heavy, both talking at once, both furiously writing equations on the genius pad. Suddenly, Walt reached over and broke the other guy's pencil. ... And then he said, 'Now! As I was saying.' On another occasion, when a good argument was going on in the next office, I saw Walt fidgeting. He listened and he stood it as long as he could. Finally, he climbed up on his desk, went over the partition, and joined the argument."

Two Leaders at North American Aviation During the 1950s

No single individual had a more profound influence on the career of Walt Evans than John R. Moore. The men met at Washington University in St. Louis in 1937 upon Moore's graduation and Dad's matriculation. Dad followed Moore to Schenectady in 1941, to Washington University in 1946, and to North American Aviation in 1948. They were different in many ways but shared a commitment to national security during the Cold War.

Washington University's McKelvey School of Engineering awarded John R. Moore and Walter R. Evans its Engineering Alumni Achievement Awards in 1988 and 1990, respectively.

> John Moore: *"Walt was one of my best friends, and I brought him from GE to Washington University and from Washington University to North American Aviation. One of Walt's major characteristics was his sense of humor."*

In 1959, Dad supported Moore's nomination for an honorary degree from Washington University with the following letter:

> *John Moore was the key driving force behind the North American Aviation team which developed and manufactured inertial guidance equipment from 1948 through his presidency of the Autonetics Division in the 1960s. This guidance equipment, used in the Minuteman and Polaris Missiles, achieves an accuracy of a fraction of a mile based upon acceleration measurement and computing alone.*
>
> *John's leadership covered recruiting, error analysis, sales as well as management. He was selected executive vice-president of North American Aviation just before the Rockwell merger, which led to Rockwell's taking over the management of the company ... John is now a vice-president of McDonnell Douglas. However, the team which he formed still dominates the Electronics Groups of Rockwell International....*

Chapter 11

The Spirule Company (1952–1980)

On March 27, 1952, Dad sent a Certificate of Business form to *The Whittier News*. On April 8, 15, 22, and 29, the following notice appeared in the classifieds: "The undersigned does hereby declare that I am conducting a mail sales business at 1706 Maple Street under the fictitious name of The Spirule Company."

And so, in a modest, 1200-square-foot residence in Whittier, California, the Spirule Company was born. Coincidentally, Dad and Mom celebrated their tenth wedding anniversary the same day.

Ironically, it had been exactly two years since Professors Wheeler, Osterbeck, and Meserve had learned from Dad that NAA had no plans to manufacture Spirules because they were "not worth $30 [for a single machine shop model] and the demand is far short of 500 [for a $0.50 stamped model.]"

On the face of it, the decision to launch a mail-order business out of their residence seemed ill-advised. In April 1952 Mom was six months pregnant and burdened with Randy and me. Dad's and Mom's parents and siblings lived two thousand miles away. Dad had demands at work and faced hard-to-meet publication deadlines for *Control-System Dynamics*.

Why, then, did he choose to assume responsibility for Spirules? In part, I believe, because he underestimated demand. Of all the feedback he received from the root-locus method, the universal desire of designers and students to plot root loci with Spirules rather than two transparent pieces of plastic pinned with a thumbtack to a "genius pad," was probably the least expected. He once remarked in a letter, "Engineers must love gimmicks." Jeff Schmidt understood the appeal of a device and deserves credit for the classic disk and arm design. Dad ultimately "closed the loop" with a business model that met demand.

Luckily for him, in Mom, he had married a woman with high energy and finely tuned time-management skills, who was fully committed to his success and content to work quietly behind the scenes. She was his executive assistant,

Mom's Office

corresponding secretary, accountant, and filing clerk in addition to being an efficient homemaker and mother to his children. Whenever anyone complimented Dad on any of his children's accomplishments, his response was always the same: "It's not me, it's Arline."

With soaring Spirule sales, the demands of running the business grew. And so did demands of home management and motherhood when, in 1954, Gary was born. They had four children ten or younger. To Mom and Dad's credit, they not only made it work, the Spirule Company had a positive impact on family cohesiveness. Initially, the burden fell exclusively on my parents. But as my siblings and I got older, we got jobs. Randy and I worked in the shop (our garage) installing eyelets and smoothing the edges of the Spirule arms with a razor blade. All four of us worked in the final assembly hall (the family room), stuffing envelopes with instruction sheets and Spirules—usually in front of the television set. And, yes, we each kept track of our time working. Dad paid us generously.

One Sunday morning, I answered the front door still in my pajamas. A bewildered engineering student stood there and asked, "Is this the Spirule Company?" He'd driven straight to our house—likely desperate on the eve of a final exam. The Spirule was available in bookstores, but for some, salvation lay closer to the source. Another student from UCLA, fed up with the challenges of mastering the tool, rewrote the lyrics to Merle Travis's song "Sixteen Tons" in parody:

9728 El Venado – The Spirule Company 1954 to 2002

Some people say a man is made out of mud
But a servo-man's made out of blizzards and blood
Blizzards and blood and a Bode plot
A moment's rest is what he ain't got.

You do sixteen tons, and what do you learn?
Slide-rule's busted and the Spirule won't turn.
Saint Peter, don't you call me, 'cause I can't go—
I owe my soul to Theta-0.

I got up one mornin' at a quarter to nine—
Picked up my Spirule and went to the salt mine.
I did sixteen tons of 181C—
The instructor said, "Well, glory be!"

THE SPIRULE COMPANY (1952-1980)

You do sixteen tons, and what do you learn?
Slide-rule's busted and the Spirule won't turn.
Saint Peter, don't you call me, for I must bleed—
I owe my soul to the lag and lead.

If you see me study, better walk away.
Or you'll never see the dawn of another day.
Read Evans with my coffee, Nixon with my meal—
If the first one don't kill me, then the last one will.

You do sixteen tons, and what do you learn?
Slide rule's busted and the Spirule won't turn.
Saint Peter, don't you call me, 'cause leave I can not—
I owe my soul to the Nyquist plot

I was born one mornin' as the root-locus flew.
Buried neck-deep in snow and a-turnin' dark blue.
Raised in the Nyquist by an old zero—
Get a grade in this course and I'll be a hero.

You do sixteen tons, and what do you learn?
Slide-rule's busted and the Spirule won't turn.
Saint Peter, don't you call me, there's too much fuss—
I owe my soul to the root-locus.

Among the thousands of orders, several stood out. In February 1959, about one year after Dad revamped the Spirule design, he received a kind letter from Professor John G. Truxal, now head of the prestigious engineering department at the Polytechnic Institute of Brooklyn. His letter reads:

Dear Mr. Evans,

Thanks very much for sending me the new model Spirule. I am very impressed with the construction of this model and I anticipate that we will be asking our students to obtain it in next fall's classes. I certainly am sold on your root-locus method implemented with the Spirule and consider this the outstanding post-war contribution to the control field.

Sincerely, John G. Truxal

Noteworthy letters need not have been written by noteworthy authors. In 1973, Mom received a letter from the Philippines. Its author, Ho Hwa Hui, asked simply: "How much does a Spirule cost?" Without hesitation, Mom mailed one out the same day, trusting he'd pay. He did—and more. His reply read:

> *You have unknowingly given me a two-fold surprise package. First, you sent the Spirule to a complete stranger thousands of miles away without even a guarantee that you'd get paid! Secondly, you are Mrs. W. R. Evans—wife of one of the most respected and prestigious control engineers in the forefront of this highly technical field. Of course, I don't have to tell you.*

Dad priced the Spirule barely above cost, aware that most of his customers were students. Orders arrived for decades, addressed to the home he moved into in August 1954: 9728 El Venado Drive, Whittier, California. That address is more than a line in a textbook—it's where my roots are. Not roots in the reckless fast lane, but roots planted firmly in the stable left half plane.

For Walter Evans, the Spirule was never just a product. It was a simple yet elegant analog computer; it enabled real-time construction of root-locus plots using the geometric properties of poles, zeros, and complex-plane symmetry.

Instructors worldwide adopted the Spirule for control systems courses; it appeared in dozens of textbooks, particularly throughout the 1960s and 1970s. From universities to defense contractors, its influence was pervasive. Between 1952 and 1986, Spirule sales reached more than 101,000 users in 50 states and 76 countries.

Control Theory Continues to Advance in the Age of Computers

For more than two decades after its invention, root locus stood at the center of control system design. It offered engineers a way to see how changes in a system—say, a feedback amplifier or a missile guidance controller—would affect its behavior. From the 1950s through the early 1970s, root locus was taught in nearly every engineering school and used daily by practitioners in aerospace, industry, and academia. But by the mid-1970s, the ground was shifting.

Computers were transforming what was possible in engineering. Instead of relying solely on hand-drawn plots and rules of thumb, engineers could now calculate precise responses of complex systems—things that would have been too messy, or outright impossible, with slide rules and graph paper. At the same time, a new generation of theorists was bringing mathematical sophistication into control theory. What emerged is often called modern control theory.

Building on this, the 1970s and 1980s saw the rise of optimal control and linear quadratic Gaussian, which used mathematical optimization to design controllers that could perform well even in the presence of noise and uncertainty. Often, the controller would work beautifully in theory. But then it would break down in the real world when conditions deviated slightly from the model. In response, a new wave of theory emerged called robust control, which aimed to build systems that could tolerate imperfections.

This transition was well underway when a young professor, John Doyle, arrived at Caltech in 1985. Professor Doyle and his collaborators introduced techniques called "μ-analysis and synthesis" and "H-infinity control." These techniques can be thought of as generalizing root locus and Bode plots to complex systems. Yet through all this, root locus never really disappeared.

In essence, the Spirule was the physical manifestation of Dad's root-locus method. It was a tool that embodied the core principle of plotting how the roots of a characteristic equation move with changes in system parameters. It transformed abstract mathematics into a designer's tool—repeatable and fast.

Because of that, it outlived the era of analog computing and slide rules. The root-locus method became embedded in MATLAB code and CAD tools. Its teaching value—as a graphical, hands-on, geometry-based approach—remains foundational in control theory instruction to this day. That is the legacy of the Spirule Company. And it is inseparable from the vision and values of Walter R. Evans.

This book has told the story of how root locus was invented. But it also sits within a larger story: how human beings have learned to tame complexity, to shape the future behavior of machines and systems. Root locus is a part of that story—one of the first and most elegant ways we learned to draw the line between stability and chaos.[34]

INTO STABILITY

Annual Sales of Spirules 1950–1987

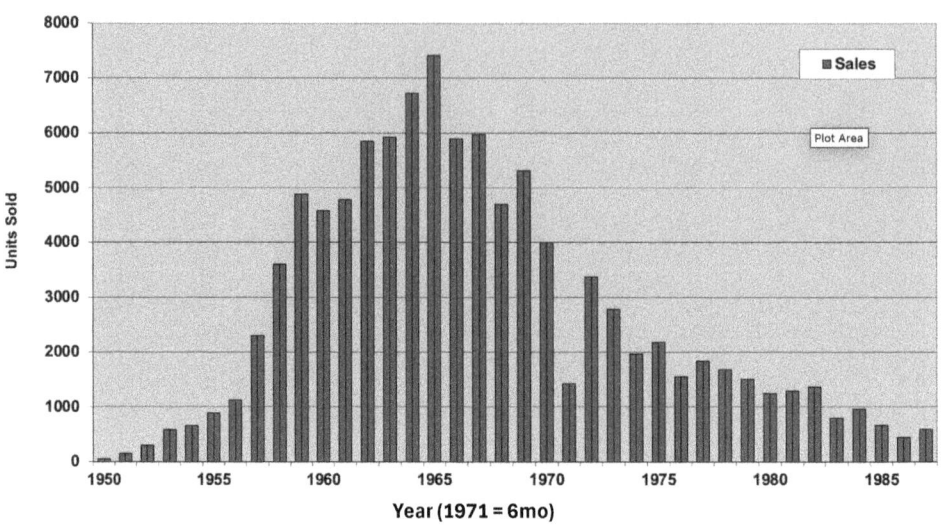

List of Countries Where Spirules Have Been Shipped

Argentina	El Salvador	Kenya	Saudi Arabia
Australia	England	Korea	Scotland
Austria	Finland	Lebanon	Singapore
Belgium	France	Libya	South Africa
Brazil	Guatemala	Mexico	Soviet Union
Canada	Greece	Mozambique	Spain
Ceylon	Honduras	The Netherlands	Sudan
Chile	Hong Kong	New Zealand	Sweden
China/Taiwan	Iceland	Nigeria	Switzerland
Columbia	India	Norway	Thailand
Costa Rica	Indonesia	Oman	Turkey
Cuba	Iran	Pakistan	Uganda
Cyprus	Iraq	Panama	United States
Czechoslovakia	Ireland	Peru	Uruguay
Denmark	Israel	Philippines	Venezuela
Dominican Republic	Italy	Poland	Wales
Ecuador	Japan	Portugal	West Germany
Egypt	Jordan	Puerto Rico	Yugoslavia

Epilogue

Arline and Walter in Whittier, CA, c. 1992

On the afternoon of Monday, June 2, 1980, I was sitting at my desk when the phone rang. I recognized my brother's voice immediately, but not its tone.

"Dad is at Presbyterian Hospital," he said, his voice cracking. "He had a stroke."

That evening, I walked into Dad's hospital room and was shocked by his appearance. His face was lifeless. My brother Randy was there. "Mom was here but has gone home for the night," he told me.

When the neurologist arrived, it was just the two of us. His report was unequivocal: Dad would never walk again or regain his speech. The stroke had destroyed a large portion of his left hemisphere, including Broca's region—the center of speech production. We left in shock.

Early the next morning, I sat with my mother in the bedroom and relayed the grim prognosis. What followed was a flurry of phone calls to friends and colleagues, breaking the news. I had always known Dad loved his work, but that night, I learned something new—his colleagues loved him, too. The pain in their voices revealed the depth of their admiration. It underscored an essential truth: The impact of his work was not just technical, but also deeply personal to those who had learned from him.

That moment of realization stayed with me for 45 years until now, leading me back to the book I began in 2002, but had set aside. Returning to the boxes stacked floor-to-ceiling in my closet, I was reminded that lack of material was never an issue, especially for the most transformative decade of his life: 1944–1954. In 1944, at 24 years old, he was just beginning to explore servomechanisms, married but childless, and earning 75 cents an hour at General Electric's Schenectady Works.

A decade later, he had authored two of the most cited papers in *AIEE Transactions*, published a textbook with McGraw-Hill, and moved into a custom-built home. He had become a father of four and a team leader at North American Aviation's Aerophysics Laboratory, the nation's preeminent developer of inertial guidance systems. His reputation was well established.

The neurologist's predictions were not far off the mark. However, thanks to her devotion, for 19 years—6,978 days—Mom devoted herself to Dad. She displayed the qualities he had espoused when he summarized a teacher's goal: to get each new class of rookies off of the bench and into the game. Dad's stroke had made him a "rookie" disabled person; Mom got him into the game. She developed a daily routine designed to minimize cognitive loss and to develop new skills to compensate for those he had lost.

Daily Routine

First thing every morning, Mom would require Dad to tell her the day of the week and verbalize the names of objects in the hall bathroom: toothbrush, sink, washcloth, and so on. There were worksheets to complete with fill-in-the-blank sentences, long division problems, and writing the names of five items belonging in a category (e.g., fruits, makes of cars).

After breakfast, there were puzzles in the morning paper to work on. Jumble Word was a favorite. At some point in the early 1980s, Mom brought out a set of colored pens and pencils and pictures of birds and animals. Dad learned to draw with his left hand. Examples of his artwork are displayed in Appendix 12.

In addition to helping Dad with cognitive exercises, Mom integrated physical exercise into his daily routine. At first, she assisted Dad in walking laps around the backyard. This exercise gave way to workout sessions on a stationary bike and daily laps at the indoor YMCA pool, with floaty rings around the limbs he could not move to keep them afloat. He had swim sessions at the Whittier YMCA. And yes, Mom recorded every lap time Dad swam, to the second. The pages fill a three-ring binder.

On September 14, 1982, a little over two years after his stroke, Mom added a new routine to his schedule. Hillcrest Congregational Church in Whittier, which the family had attended since its founding in 1955, began a twice-a-week, two-hour-a-day program called Senior Caring Fellowship, under the leadership of Rev. William Meyer.

Mom dropped him off there twice a week for more than seventeen years. Consistent with her record-keeping nature, she made an entry in a notebook for every visit—more than 1,700 entries on about thirty two-sided sheets. In a small but very readable script, she fit sixty sessions, or about six months, on two sides of one sheet.

Mom also kept a daily diary of Dad's activities for ten years, between 1980 and 1990. She filled a fifty-page notebook that documented every visit with specialist Dr. Lester "Les" Harris from Whittier College. (Harris directed the college's Speech and Hearing Clinic from 1966 to 1998.)

Chess

Many people were surprised to learn that Dad played chess. Even fewer people knew about his grandmother, Eveline Allen Burgess. Her father, James X Allen (a field surgeon for the Union Army), taught her to play chess during her school lunch hours. Years later, Eveline accompanied her husband, Samuel Burgess, annually to New York. He was a buyer for a St. Louis stationery store.

On a 1906 trip, she challenged the reigning women's chess champion, won the match, and was declared to be the new US Women's Chess Champion. Mom did not play chess, but Dad's colleagues chipped in to get him a Sargon chess computer to use at home. Dad had many opportunities to play chess at Senior Caring Fellowship. Several of the volunteers learned how to play the game in order to play with him. One, Charlotte Verdery Lightfoot, would play for fifteen years. One young man, Bill Ford, often fell behind. When a game became one-sided in Dad's favor, Dad would simply turn the board around on his next move and offer to play from the weaker position.

Poppy

Dad had five grandchildren: my niece and nephew, my daughter, and my two sons. They knew him only after his stroke. Yet he was very much a presence in their lives because he was always included. His body language made it clear he understood everything being said. If he wanted to "speak," he signaled with his left hand. Given a pencil and paper, he would write a word or phrase. We'd ask follow-up questions, and usually we could figure out what he meant. It worked best when his written word gave us a solid start and we followed up with good guesses and approximate answers. In that sense, it was a lot like creating a root-locus plot.

Some traits that stood out after the stroke—and often drew comments—were his smile, his thumbs-up sign, and the iron grip of his left hand. That's how his grandchildren remember "Poppy."

Dad's drawings with his left hand are remarkable. Because of the brain's cross-wiring, the left hand is controlled by the brain's right hemisphere. Over 250 of his drawings are organized by date in a five-volume portfolio. About half were based on bird and mammal stamp images. The other half came from his memories of the farm or from his imagination.

A Legacy of Love of Life

Five years after the stroke, William Meyer, Director of Senior Caring Fellowship, told Mom, "All the participants are deteriorating—except Walter. He is improving." When Dad heard that, he wrote, "I'm a winner." Indeed you were, Dad. What was most unexpected and thus what stays with me about Dad after his stroke was his acceptance of his condition and his positive attitude. The man whose lifeless face stunned me in June 1980 found new ways to express himself, to be present, and to live fully. That love of life informed his sense of humor, his engineering, and his art—a legacy I've tried to honor here.

EPILOGUE

A Legacy of Understanding

One of the themes running through this story is that the ways engineers think about problems shape their ability to solve them creatively. A breakthrough doesn't come from memorizing steps. It comes from seeing the heart of a problem clearly, from the right angle, until the answer reveals itself. Mathematics becomes the language used to express that clarity—not its source. Richard Feynman was that kind of thinker in physics. Dad was that kind of thinker, too.

I came to that understanding in the mid-1960s when, as a student, I encountered Feynman. Dad would drive from Whittier to Caltech to join study sessions. He had his own set of The Feynman Lectures on Physics. He was in his element. Feynman's explanations of subjects like refraction weren't textbook. He didn't start with equations; he started with atoms. An atom doesn't know what an index of refraction is. It just behaves. The equations came later.

Dad's physical understanding was the same. Jeff Schmidt once noted that Dad was likely the first to express a term in inverse form, not because he was trying to innovate, but because the answer became obvious that way. He flipped the equation to match how he already saw the system. What connected these two men—Feynman and Evans—was not just brilliance, but also clarity. They both had the ability to flip a problem over like a puzzle piece until it fits. They both were able to ask: Is this problem easier to solve as miles per gallon or gallons per mile? To find the frame that makes the solution visible.

There was one more quiet link between them: They both married women named Arline. Feynman's Arline died heartbreakingly young. Dad's Arline was luckier. She stood beside him through the long years when he could no longer speak. She turned the pages; he turned the problems over in his mind.

Even after his stroke in 1980, Dad's desire to understand never left him. Mom read aloud Ralph Leighton's books about Feynman. When the Challenger disaster occurred, he followed Feynman's analysis closely, especially the blunt appendix that laid bare the human failures behind the technical ones. This story, then, is not only technical history. It's a reminder that the clearest path to a solution—and the deepest kind of loyalty—often begins with how, and with whom, we choose to see the world. And in that choice lies the truest legacy of all—not the recognition we earn, but the understanding we leave behind.

A Legacy of Grace

I conclude with words spoken at Dad's 1999 memorial service by Dr. Edward H. Bloomfield—a friend, our pastor, and an officiant at Evans family baptisms, weddings, and memorial services for more than five decades. Ed was a large man with a hearty laugh, yet the word that best described him was gravitas. After he spoke about Dad's character, his sense of humor, and Mom's response to Dad's stroke, Ed turned his attention to their partnership.

> *It was a privilege and an honor to have known Walter R. Evans. He was a gifted and extremely intelligent man with a great sense of humor. He was humble and gentle and never sought the limelight for himself. Walt, of course, was also a mentor—professionally speaking. But we are all beneficiaries of his mentorship of how to face life courageously, no matter what hand is dealt to us.*
>
> *Now, here I cannot help but say a few words about Arline. My dear, for me, you are truly wonder woman. What mentors and role models you and Walt are for all of us. You lived your marriage vows day-by-day, for better or worse, in sickness and in health. And as a team, Walt and Arline have demonstrated, as clearly as any couple whom I have ever known, that marriage is not for the faint-hearted. … In a time of great hardship, Walt and Arline taught all of us … what life is really all about. And that ultimately, what really counts is constancy, faithfulness, and love.*
>
> *So let us leave today with grateful hearts that we have the privilege of knowing Walter R. Evans. And that we are the beneficiaries of his legacy as teacher, as mentor, and as (an) example of grace.*
>
> *Amen*

EPILOGUE

Walter and Arline c. 1985, Whittier, CA
"I'm a winner!"

CHAPTER 4: THE GRAPHICS ANALYSIS PAPER

Robert Doherty
Founder of General Electric's advanced engineering program

Paul Profos
Swiss engineer who wrote a graphical analysis paper for the *Sulzer Review*

CHAPTER 7: THE INVENTION OF THE SPIRULE

Jeff Schmidt
Conceived of Spirule and built the first in 1948

DeWitt Lyon
Coined the name: Spiral + Slide Rule

CHAPTER 8: THE ROOT LOCUS PAPER

Gordon S. Brown

Chair of *AIEE Transactions* Review Committee in 1949

Williams Bollay

His December 1950 Wright Bros. Lecture gave Root Locus national exposure

CHAPTER 9: THE TEXTBOOK

Fred Terman

Founder of McGraw-Hill series and reviewer of *Control-System Dynamics*

George Thaler

Author of first McGraw-Hill textbook to devote a chapter to Root Locus

Cardinals by Walter R. Evans, January 1 1987, Portfolio #226, Vol. 3

Appendices

Appendix 1: The Root-Locus Method of Walter R. Evans

Appendix 2: New Challenges for Engineers (1945)

Appendix 3: The General Electric Years (1941–1946)

Appendix 4: The "Root Locus Idea" (1949)

Appendix 5: The Invention of the Spirule (1948–1951)

Appendix 6: Historical Spirule Documents (1948–1951)

Appendix 7: Profiles of Prominent Engineers

Appendix 8: Correspondence from Autonetics Colleagues

Appendix 9: Walter R. Evans Biographical Information

Appendix 10: The Quotable Walter R. Evans

Appendix 11: Pinball: Polynomial Factoring with Root Locus

Appendix 12: The Artwork of Walter R. Evans

Appendix 1

The Root-Locus Method of Walter R. Evans

This article on the root-locus method of Walter R. Evans was prepared for this book in 2004 by Professor Robert H. Cannon Jr.

At the heart of any automatic control system's design is what that system's natural dynamic behavior will be. How stable will it be and how quickly and well damped will its natural motions be following any disturbance? (Second in order of importance—but also, of course, central—is how well the controlled system will follow commands.)

The Characteristic Equation (CE) for a Linear System

Suppose we have a set of linear differential equations that completely describe the dynamics of a physical system. To determine the character of the natural dynamic behavior of which the system is capable, we represent that the behavior for each variable (y) in the equations will be of the form $y = Ye^{st}$. This assumption always works simply because, when you differentiate each variable, you get back the original form multiplied by s:

[1] $dy/dt = sYe^{st}$

With this substitution, each differential equation is converted to an algebraic one. Then by combining the resulting algebraic equations to eliminate all the capital letters (Y's), you get one simple algebraic equation in s, of the form:

[2] $s^5 + A_4 s^4 + A_3 s^3 + A_2 s^2 + A_1 s + A_0 = 0$

This is the system's CE. The next step is to factor it:

[3] $(s + α)(s + β)(s + σ - jω)(s + σ + jω)(S + γ) = 0$

When the roots α, β, and so on are each put into the original assumed motion for y, they connote quantitatively the character of one of the natural motions the system can have. For example:

[4] $y = Y_1 e^{st}$

Any of the Greek letters might be zero, connoting that a root has the value 0. For the complex pair of roots, the value of ω indicates the frequency of a sinusoidal motion and the value of σ, the motion's rate of damping.

Factoring the CE [2]

This is the crux of the matter: determining each root of the CE. For a control system, this equation can be written so that the part of the equation that contains the control variables (which are at the disposal of the designer) are separated from the rest of the CE. For example, the CE could turn out to be of the form:

[5] $s(s+b)(s^2 + Bs + C) + K(s+a) = 0$

where K and a are the control variables that can be selected. When multiplied out, [5] would give back the original CE, [2].

This equation can be written in the powerful Evans form:

[6] $[s(s+b)(s^2 + Bs + C)] / (s + a) = -K$

Next, the values of s that make this equation correct for each value of K are in fact roots of the system's CE for that K. Plotted in the complex plane (i.e., the "s-plane": the y-axis is the imaginary part of s and the x-axis is the real part of s), these values of s form the locus of roots. Hence the name "root locus."

The Root-Locus Plot

The way the plot is constructed begins with finding the roots for the value K = 0. Then, *in principle*, one must "search" for all the other points in the s-plane that satisfy Equation [6]. What makes Evans's reformulation of [5] into [6] so powerful? Following Evans's incisive instinct, the search for roots is reduced to two simple parts.

Part 1 is finding the loci of all the locations that simply meet the total phase angle 180 degrees that corresponds to the sign in [6]. For a trial point, the angles of the vectors are summed. If the total is 180 degrees, the point will satisfy [6]. The locus of all these points in the s-plane is the "root locus" of Equation [6].

Part 2 is establishing the magnitude of K at important points of the locus. For example, we might determine the value of K at which the roots pass from a stable condition to an unstable one (or vice versa)—that is, the value of K when the roots cross the imaginary axis in the s-plane and the real part of s transitions from a negative, decaying exponential to a positive, growing exponential. Stability requires all roots to lie in the left-half-plane.

Another important value of K on the root locus occurs where a desired level of damping is achieved; this is defined as a line in the s-plane crossing the real axis at specified value for re(S).

As noted, all of the above can be done in seconds today on a common powerful laptop computer. But in 1948 there were no digital computers. So, Evans co-invented [Author's note: with Jeff Schmidt, as described in Chapter 7] a plastic device for making the above computations, in minutes, without electricity!

The Spirule: For Plotting the Locus of Roots Quickly and Accurately

Part 1. The locus of roots

To make a Spirule, you start with a circular plastic disk of diameter 4½ inches. Using radial lines, mark directions 0°, 90°, 180°, and 270°. Next, make a tick mark for each degree on the circumference and label every 10 degrees with its value.

Now drill a hole in the center of the disk for a small circular clamp with a hole in its center. Clamp to the protractor and atop it a straight-edge ruler ("arm") that extends out 9¼ inches from the center of the hole. Give the clamp just enough friction to permit setting the arm at any angle and having it hold well. Finally, give the bottom edge of the clamp a sharp edge so that when you push down on it, it holds fast to the paper.

You are now ready to carry out Part 1 of the process for plotting the roots of any CE. Begin by putting the CE in the form of Equation [6] and plotting the roots for K = 0 as Xs on the s-plane. Next, choose a place where, for some value of K, a root is likely to be found. This is trial and error. However, it can be fast because you can precede the Spirule's calculations by using the quick rules Evans developed to sketch where loci are likely to go. Each trial takes seconds.

Thus, to check whether a given spot in the s-plane yields a 180° product of the vectors, simply (i) draw a straight line horizontally to the left from the spot, (ii) set the straight edge to 0° and place the center of the Spirule's hole over that spot, and (iii) add up the angles between the horizontal and each X by swinging the straight edge, holding the disk down while going from horizontal to X, and then letting the disk stick to the arm when swinging back. It's really a quick process to find the angle at the point being measured. Mark the value of the product on the plot. Then try another point until you find a 180° point.

Three Sketching Rules: For the Loci of Roots of a CE

A presentation of Evans's three sketching rules is found in many sources (e.g., *Dynamics of Physical Systems*, Dover Publications, 1967). These rules provide a methodology to perform Part 1 quickly. Key to speed is the acceptability of approximate results initially followed by refined loci locations only where required.

Part 2. The magnitude of K at any point on the locus of roots

With the Spirule's center sitting on any point on a locus, it is entirely straightforward to obtain the value of K by measuring the lengths of the vectors (using the scale on the straight edge) and then multiplying the lengths together, as Equation [6] indicates. In 1948, the multiplications could be done by using a slide rule or by entering the lengths into a Marchant calculator and punching "multiply." Time-consuming!

Evans invented a more elegant way. You use the logarithmic spiral on the arm of the Spirule. You have a slide rule, but you don't need to set any numbers into it. Instead, you set in the length of each vector graphically.

To insert the length of a vector, put the center hole on its point and the reference line (R) on its tail, and then—holding the disk fixed—rotate the Spirule until its curve is on the vector's tail. After you've done this for all the vectors you want to multiply together, look at the output arrow and read their product from its scale.

You now have the locus of roots plotted with the values of K at the points of most interest. The first step in designing a good control system is finished: It will be stable, with K selected to provide a good margin of stability despite possible changes in physical parameters. Its natural motions will be acceptably quick, and its oscillations will be acceptably well damped.

Appendix 2

New Challenges for Engineers (1945)

by Walter R. Evans, Associate Editor

Published in Schenectady Engineering Council Bulletin, *Vol. III, No. 3, November 1945*

The challenge of war is definite: to win. The engineers may all be proud of their record of development of amazing military equipment. One may well ponder in retrospect what were the factors which contributed to this achievement. Certainly, one of them is the mere fact that the purpose of our efforts was clear-cut. The profession is now free to accept new responsibilities. But the challenges of peace are not so definite. We need the stimulus of discussion to realize the many problems which await solutions. The purpose of this article is to set forth in very general terms three problems, which new devices may help solve: maintaining full employment, stimulating long-range scientific thinking, and speeding the processes of democracy.

The challenge of reconversion is very important to the economic welfare of our nation. The key man in this problem is the engineer for he must have the new designs ready when the production facilities are available. Fortunately, the demand for pre-war products not made during war will maintain employment for a few short years. But for the designs to be ready then, the ideas must be conceived and the application developed now.

The application of wartime developments to peace provides a fertile field for imagination as shown in the following examples. How can the know-how of Radar be more useful than just the obvious direct use for ship navigation. The B-29 computer as such looks like a total loss to anyone merely expecting to use the same thing somewhere else, but the ingenuity shown in many of its features spur on the designer who starts to feel that his problem cannot be solved.

The challenge of the future is in some respect more inspiring than that of the present. Let us consider the possible advances of science which we would like to achieve even though the exact means to be used are perhaps quite vague. Dr. Vannevar Bush has provided an excellent example with his conception of "Memex,"* the automatic aid to the memory. He points out how man's accumulation of knowledge is far beyond what any one man can comprehend. But removing the burden of inefficient library research and organization of notes would be a big help.

Consider a desk in which microfilm recorded any hand note, page of a book, or picture which you might want to recall. A coding system using the technique of stenotype could identify this film. Automatic selection mechanism would then flash this film back on a screen at the press of a button. A chain of associated ideas might be linked by the code and flashed back in the original sequence. A person could make up duplicates to exchange with his friends. The mind would thus be free for more creative thinking with less fatigue from routine. Certainly, all engineers have at one time or another been participants in a "hare-brained idea session."

Admittedly most of the devices jokingly suggested are more appropriate for cartoons than for manufacturers' catalogues. But for those occasional times that a basically sound idea is developed, the sessions are worthwhile. Besides that, we all apparently need some such stimulus to our imagination to break the channels of our thought from the job immediately before us.

The challenge of making democracy work more effectively is one which the engineer may aid in far more than his capacity as one citizen. The radio and the airplane have helped get the views of the leaders to the people but consider how uncertain is the means by which the leaders in turn learn the will of the people. The present crisis of strikes throughout the nation illustrates the problem. The union demands for maintaining the same "take home" pay for decreased hours of work is more than just the usual give

---

* In 1945, Bush authored the article "As We May Think" in *The Atlantic Monthly* in which he first proposed his idea of the Memex machine. This machine was designed to help people sort through the enormous amount of published information available throughout the world. This description, which was written about thirty years before the invention of the personal computer and fifty years before the birth of the public World Wide Web, lays out the notion of the modern link. The Memex was to be a storage and retrieval device using microfilm that would consist of a desk with viewing screens, a keyboard, selection buttons and levers, and microfilm storage. The machine would augment human memory by allowing the user to make links, or "associative trails," between documents. Bush proposed the notion of blocks of text joined by links and introduced the terms "links," "linkages," "trails," and "Web" through his descriptions of a new type of textuality. Bush's article greatly influenced the creators of what we know as "hypertext" and how we use the internet today.

and take with individual companies. The settlement requires a decision of national policy because it affects the value of the dollar.

Yet how are the representatives of the people elected many months ago going to know the wishes of the people on such an issue? Consider the basic unit of democracy: the discussion meeting. All must have felt the growing feeling of despair as time rolls by with one or two persons quibbling over small points. Time is also wasted by forum speakers who are supposed to set forth the real issues but too often dwell at length on points already accepted.

What if the audience's opinion both as to interest and agreement could be measured instantaneously and revealed by large meters on the stage? The result could be easily attained by having each person in the audience register his opinion by turning the knob of a potentiometer. Connecting these potentiometers together in any one of several standard circuits could be used to actuate the meters.

Note that the individual vote would be secret and would not interrupt the discussion. Think then of how the radio audience reaction to a program could be registered back at the studio. These ideas are not new, but perhaps the time is now at hand for putting them into service. Such a development would test not only the engineer's technical ability, but also his ability to understand people.

The challenges here presented are merely the result of one engineer's reading and discussion with friends. The vastly greater number of challenges and the more detailed nature of those mentioned which must be in the minds of the many engineers in this area particularly is difficult to imagine. The setting forth of these ideas before the entire group should crystallize the challenges and awaken much dormant ability. Then we might consider how we can combine our efforts more effectively in meeting the challenges.

You are therefore strongly urged to send in your ideas on the challenges facing the engineer. The *Bulletin* has set up a new feature, "The Voice of the Engineer," as a means to make such widespread presentation of ideas possible. You will find the question for this month to be "What are the new challenges facing the Engineering profession?"

Appendix 3

The General Electric Years (1941–1946)

Dear Greg,

I wish that I had more information to send to you, but I hope that this will help you and your project. However, getting it together has reminded me of the good times I had with your father and the good memories that I have of those days.

The bicycles were an important way to circulate in Schenectady during World War II, definitely an advantage over the overcrowded, overworked, and under-reliable buses. I believe that Walt and I were among the many who rode bicycles to work year round for at least three years. Winters provided the challenge, with temperatures occasionally dropping below –20 degrees and snow that most of the time was so deep that snowbanks beside shoveled driveways were higher than the cars. When I tell people that anything less than 100 inches of snowfall was considered to be an open winter, I'm not really exaggerating.

Actually, the main problem in winter was dressing for the occasion. Snowplows kept the side streets under control, and householders were conscientious about shoveling their walks. Even ice was a manageable problem because you had to be balanced anyway and you knew better than to make any quick moves. Moreover, you were well padded. Getting to the office meant peeling off at least three and probably as many as five layers of clothing, accompanied by the heaviest wool-lined mittens (no gloves), at least one scarf, and near-Eskimo headgear. Since legs had to be flexible, they were less well protected and sometimes required extra warm-up time at the office.

I've enclosed some items of literature on the Advanced Course because it is the key ingredient in my relationship with your father and also in his distinguished career.

These pieces give you a reasonably clear idea of the sequence of the courses in the three-year program. Entrance depended on the results of a personal interview and the grade in an examination. I have heard of only one person in all the years since 1922 who aced those exams. (Classes started in September and were identified by the year of their May completion, so the first A Class was completed in 1923.) The average grades among bright new engineering graduates were generally below 35, and scores in the 40s and 50s were considered excellent. (The course graduates who supervised the classes and devised the exams considered it a matter of pride and necessity to make certain that the younger generations scored no better than their predecessors.)

The class "supervisors" were picked from the current students and, for the C Classes, from recent graduates. They were responsible for following the general list of topics, ranging from thermodynamics and kinetics to elasticity and electromagnetism, selecting the homework problems and then reviewing and grading the results. Occasionally, they would include a surprise quiz on a technical subject not necessarily related to the current lectures. The lectures came from no textbook but were delivered mainly by practicing engineers in the company, probably but not always former Advanced Course members. The class supervisors also gave lectures on the basic topics, sometimes to their own class, sometimes to one of the other classes.

This exposed the class members to a large number of the leading engineers in the company, building some early networking. It also provided some practice in note taking, since solving the assigned problems depended entirely on what you could record or remember from a specific lecturer. Lecture styles varied, ranging from Philip L. Alger, an electric motor authority who spoke so softly and rapidly that notes and comprehension were almost impossible, to Gordon Carter, a transmission line expert who spoke both clearly and slowly (in a Virginian's accent) but was covering the material in giant leaps. Then, for pure mathematics, there were Dr. Hillel Poritsky, who could lose you quickly even in familiar territory, and Dr. Gabriel Kron, the expert in tensor analysis who tried to make an arcane subject clear and whose answers, according to rumors, were either off by a factor of two or had the wrong sign. However, in between, there was some important and fundamental wisdom imparted.

The point that the written articles do not cover is the degree of competitive difficulty involved. Less than half of the members who completed the first year's A Class would continue in the B and C Classes. Part of the attrition was by personal choice, but much of it was from non-selection by the staff. Performance and interest were the important factors.

Class members were holding full-time jobs during their time of participation. The only time away from those jobs during working hours was for the four-hour class period once a week. The problems were solved at home and, for the A Class, could require between 20 and 25 hours a week. Most of this time was needed to get some understanding of the basic principles involved and to express them correctly mathematically. The rest of the time went into trying to solve the resulting mathematical expressions, including simply keeping track of all the terms in the equations.

During the time Walt and I were taking the classes, factory schedules included significant overtime even before Pearl Harbor. This meant six days a week, sometimes 12 hours a day. There wasn't much time left for homework and/or families, and Advanced Course wives had to be very patient and understanding.

Incidentally, even Pearl Harbor Day had special Advanced Course significance. It's true that you can remember exactly where you were when you heard the news. On that Sunday afternoon, P. L. Alger (the motor authority) had invited all A Class members to his house for a reception; and we were there when the news broke.

The class photos that I've included show our progression through the three years of the program. Walter and I were in Harold Stocking's A Class, which was one of two running simultaneously in 1941–42. We were in Harold Chestnut's Electrical B Class and then Frank Olney's Combined C Class.

When I mentioned full-time jobs, I could have said that most members of the A Classes were new to the company and were in various parts of the factory testing the products. The Test Program was the traditional entry-level method for engineering college graduates, and for the most part they had a series of three-month assignments. They were on the hourly payroll; and I suspect that Walt, like me, had been hired at 75¢ per hour and then informed before he reported that it had been raised to the handsome amount of 85¢ per hour. Arline can probably confirm this.

The members who were accepted for the B and C Classes were transferred to the payroll of the Advanced Course and became salaried employees assigned for six months at a time to various engineering departments. There were usually more assignment openings than class members;

so, assignments for the most part were made by mutual agreement. Advanced Course members could thus begin to move toward their specific areas of interest.

I don't remember what Walt's working assignments were during our 1942–43 B Class year. Draft deferments were beginning to be a concern, but the General Electric influence was still strong enough to persuade local draft boards to leave young GE engineers untouched, especially when they were working on products clearly used in the country's defense.

This was true for the Advanced Course rotating assignments in engineering offices. It was true for a while for the class supervisors, who had full-time jobs directing the classes. When Walt and I began our simultaneous class supervisor posts in the summer of 1943, the Company believed that we were protected from the draft. However, early in 1944 I believe, the situation became critical; and the class supervisors were reassigned themselves to engineering offices. Both Walt and I landed in the Aircraft Ordnance section of the Aircraft and Ordnance Systems Department.

Walt was assigned to Harley Bixler, who was trying to improve the effectiveness of the machine gun sights for the new B-20 bomber. The guns in the B-29 system were controlled remotely from sights that the gunners were supposed to keep aimed directly at the oncoming targets. A set of ingenious mechanical computers connected to selsyn devices produced additional signals to compensate for the range to the target and the angle of the guns, for the windage caused by the plane's airspeed, and important for the rate at which the gunner had to swing the sight to track the target. The sight contained two gyroscopes, and the force necessary to cause the gyroscopes to precess to follow the sight could be measured by the current in the electromagnets moving the gyros. This system had some limitations, and when Walt arrived the group was working on a different approach.

The new idea was to let the gunner give a signal when he was starting to track a target within range, and the gyroscopes which had been rigidly fastened to the sight would be freed to remain stationary in space. A short time later, they would be clamped in their new position relative to the original one, and the change in position would be converted into an electrical signal telling the guns how much to lead the sight. Walt was involved in some of the design decisions. I'm not certain whether or not he went to Eglin Field for any of the field tests.

My own assignment, incidentally, was to design a tester to be used in the field to check the operation of the modified computer system to accommodate the new sight.

We stayed in A&OS until about the summer of 1945. Meanwhile, in 1944, there was the matter of the C classes we were taking and the B classes we were supervising. I think that the C classes were simply terminated because it was spring, but the A and B classes were continued on a sort of correspondence basis until their scheduled ending.

The result was that Walt and I spent quite a few evenings together in the otherwise deserted Advanced Course office, preparing material to be sent to our students while we carried on with our day jobs on B-20 fire control. It's probable that we looked over some homework, although my recollection is peculiarly hazy on that point.

One thing that I do remember—and the point of this rambling narrative—is that during one of those evenings Walt began to read a paper by a Swiss (I believe) mathematician on a different way to look at the roots of a polynomial equation. Walt was intrigued by it, spent more time looking at it and thinking about it, and finally developed an interest in applying it.

We had been exposed to feedback theory in Advanced Course classes, Nyquist diagrams were in vogue, and our current assignments brought us into contact with servo systems. For whatever reason, this was the period when Walt caught the bug that led eventually to his root locus analyses and the Spirule. He had the insight, determination, and ability to take the initial concept to a higher theoretical and practical level.

I just happened to be in the same room with him when the bug first bit him, but I remember his comment about an interesting analysis.

Just for the historical record, we and the other class supervisors kept on with our defense-associated assignments until the war was clearly ending. There were no Advanced Course classes for the 1944–45 year. In the summer of 1945, I went back to the Advanced Course program and became its administrator from 1946 to 1949. Unfortunately, I don't remember what Walt's next assignment was nor exactly when he left GE. Fortunately, we were able to stay in touch through the years.

Anyway, I hope that somewhere in this long recollection you can find something that will help you with your project.

With my very best wishes, Gordon Walter

"The Root Locus Idea" Letter: June 1949

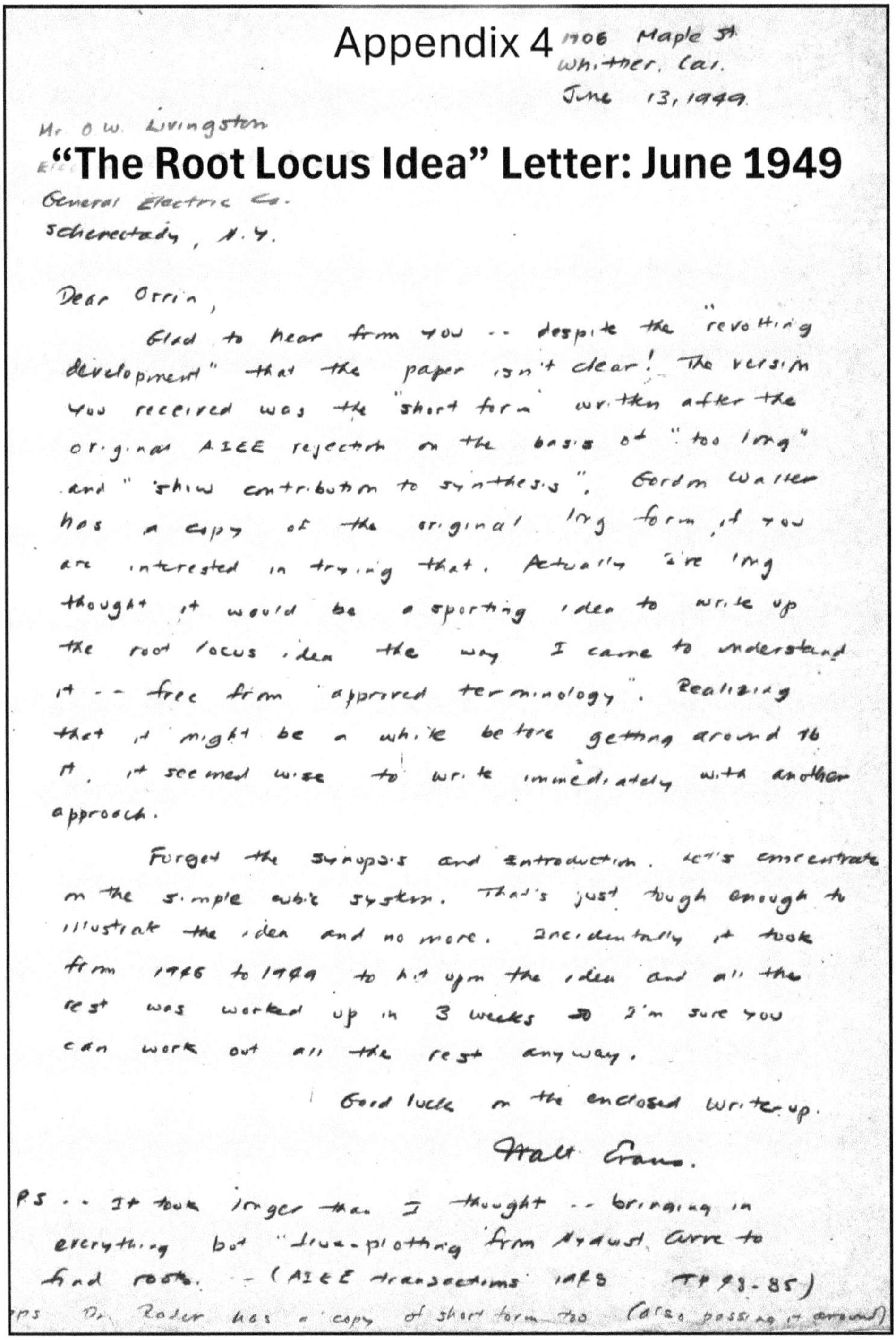

[Handwritten letter reproduced:]

1906 Maple St.
Whittier, Cal.
June 13, 1949

Mr. O. W. Livingston
General Electric Co.
Schenectady, N.Y.

Dear Orrin,

Glad to hear from you -- despite the "revolting development" that the paper isn't clear! The version you received was the "short form" written after the original AIEE rejection on the basis of "too long" and "slim contribution to synthesis". Gordon Walter has a copy of the original long form, if you are interested in trying that. Actually I've long thought it would be a sporting idea to write up the root locus idea the way I came to understand it -- free from "approved terminology". Realizing that it might be a while before getting around to it, it seemed wise to write immediately with another approach.

Forget the synopsis and introduction. Let's concentrate on the simple cubic system. That's just tough enough to illustrate the idea and no more. Incidentally it took from 1946 to 1949 to hit upon the idea and all the rest was worked up in 3 weeks so I'm sure you can work out all the rest anyway.

Good luck on the enclosed write-up.

Walt Evans.

P.S. -- It took longer than I thought -- bringing in everything but "live-plotting" from Nyquist curve to find roots. — (AIEE transactions 1948 TP 48-85)

[P.P.S.] Dr. Roser has a copy of short form too (also passing it around)

Orrin – I've long thought it would be a sporting idea to write up root locus the way I came to understand it.

… Let's concentrate on the simple cubic system. That's just tough enough to illustrate the idea. … Walt

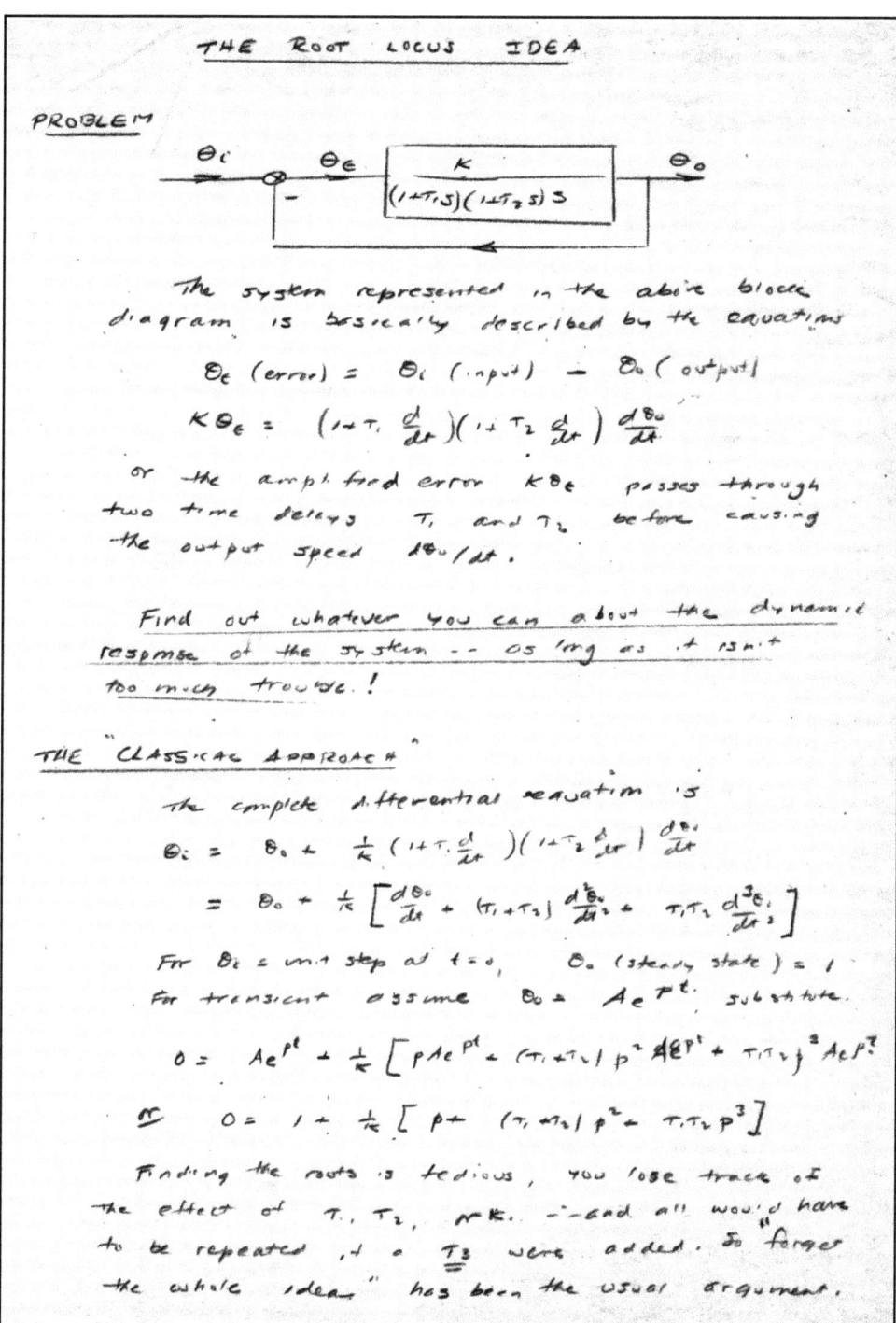

The Root Locus Idea

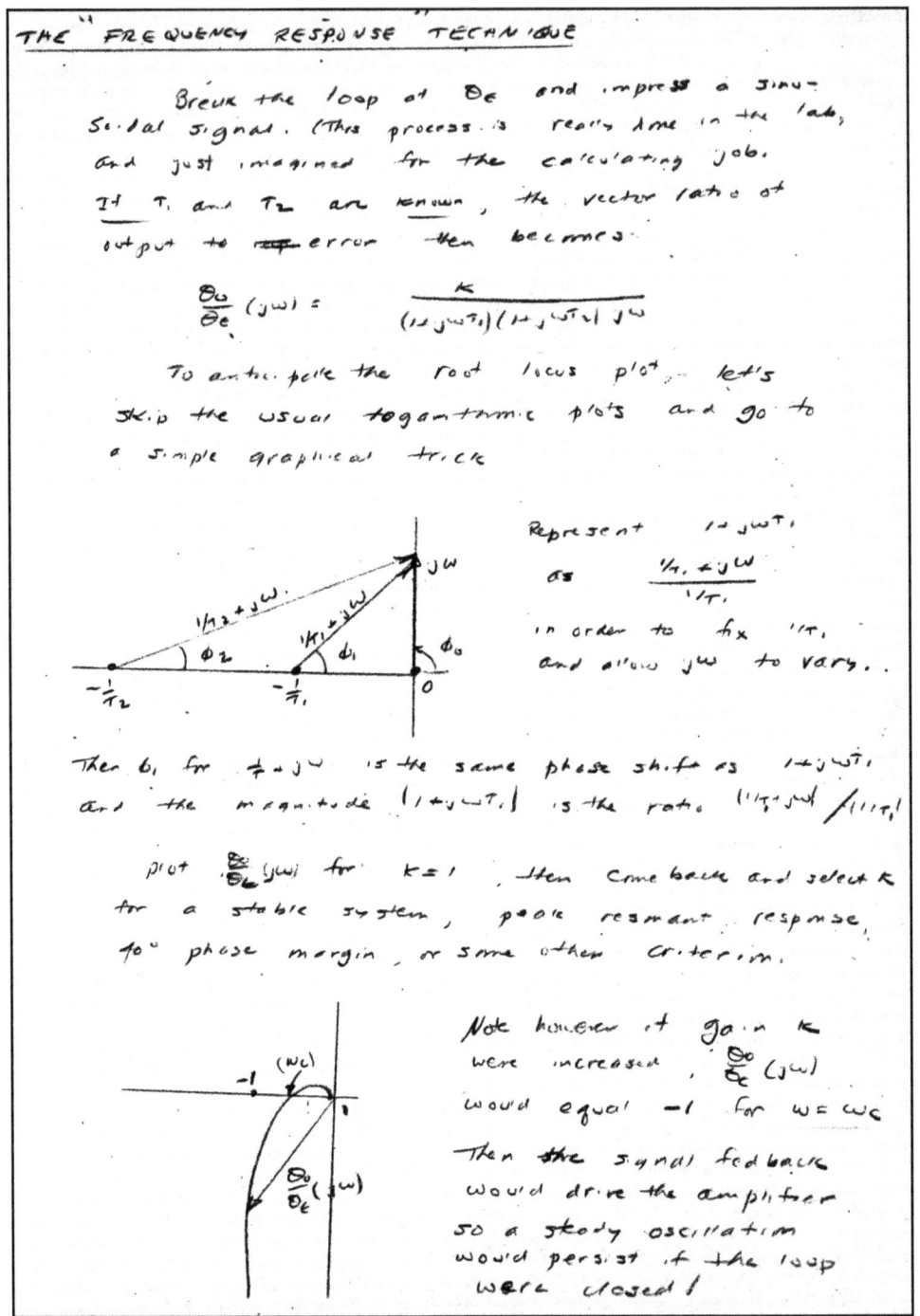

The "Frequency Response" Technique

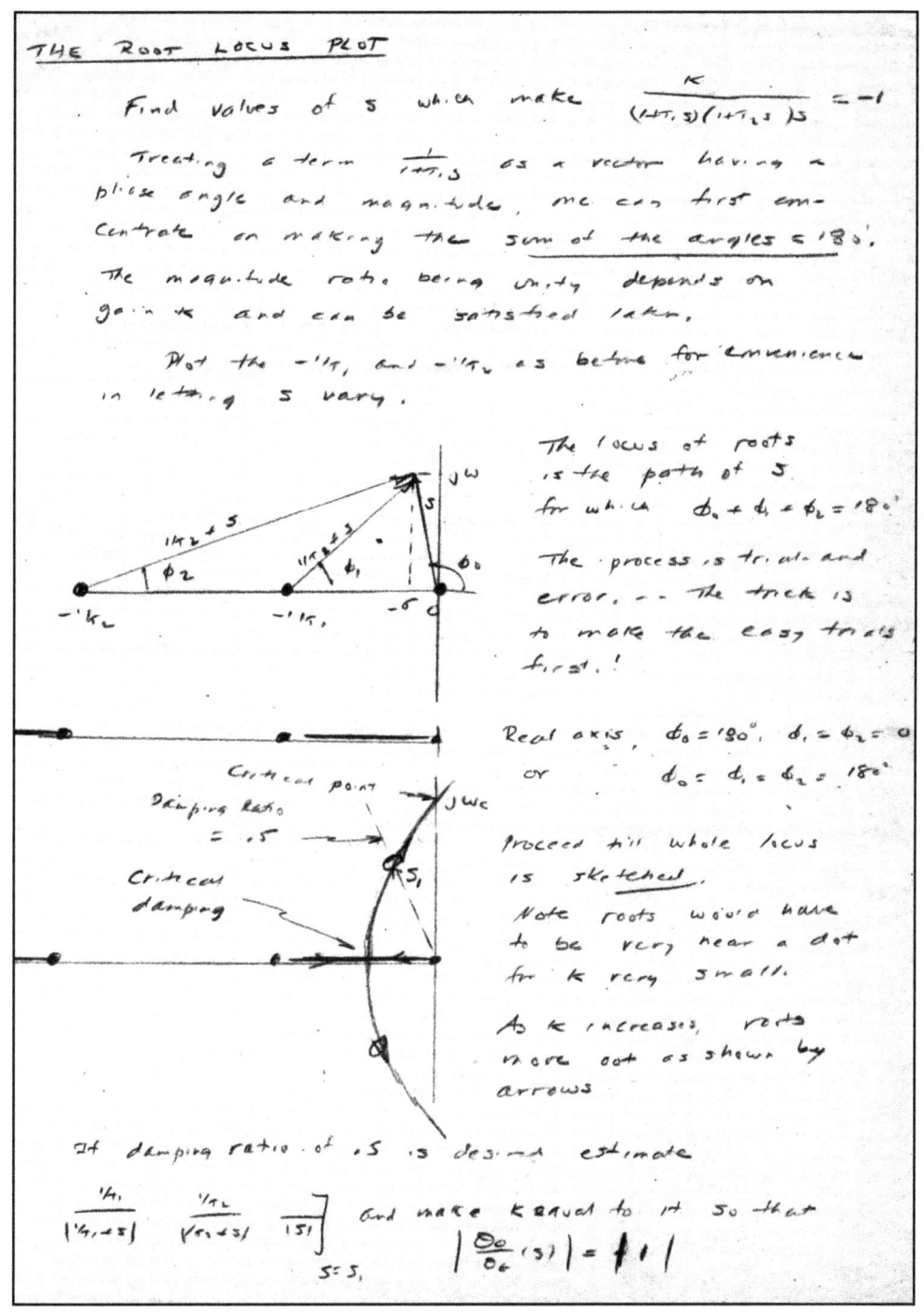

The Root-Locus Plot

ROOTS — THEIR LIVES AND HABITATS.

We just had one! — jω was a root of the differential for gain k such that the Nyquist curve went through -1.

What does + mean? jω being a root means that $e^{j\omega t}$ or $\sin\omega t$ is a signal which can "chase itself around the loop" — no driving function is required.

If gain were decreased the Nyquist curve would fall short. — but what would the system do? — oscillate with gradual damping. This raises the question, if $\frac{1}{1+j\omega\tau}$ describes what happens to a sine wave $e^{j\omega t}$, what happens to a damped sine wave $e^{-\sigma t}e^{j\omega t}$ passing through the same delay? (Incidentally, if we can swallow the idea that $5.71828^{\sqrt{-1}\,\omega t}$ represents a sine wave, it's perhaps not too much to go along with the next step that $e^{(-\sigma+j\omega)t}$ represents a <u>damped</u> sine wave!)

Let's try it on the old favorite, an RL circuit.

$$v = iR + L\frac{di}{dt}$$

But $v = V e^{(-\sigma+j\omega)t}$, so assume $i = I e^{(-\sigma+j\omega)t}$

Then $V e^{(-\sigma+j\omega)t} = I e^{(-\sigma+j\omega)t} R + L(-\sigma+j\omega) e^{(-\sigma+j\omega)t}$

$$\frac{v}{i} = \frac{I e^{(-\sigma+j\omega)t}}{V e^{(-\sigma+j\omega)t}} = \frac{1}{R + L(-\sigma+j\omega)}$$

This by me is amazing! — The same old impedance shows up with $(-\sigma+j\omega)$ in place of $j\omega$.

Roots are thus in general complex numbers $s = -\sigma + j\omega$ which represent damped sine waves which can chase themselves around a loop. Mathematically this means that $\frac{E_o}{E_e}(s)$ must $= -1$.

And $\frac{1}{1+\tau s}$ is a vector representing transfer function for <u>damped</u> sine wave.

Roots — Their Lives and Habitats

Appendix 5

The Invention of the Spirule (1948–1951)*

One of the advantages of being old is that people who might question your memory are either dead or also have questionable memories. With that said, let me [Jeff Schmidt] recall a few things about Walt Evans.

In the summer of 1948, I was working on the NATIV missile test program in New Mexico. Ray Curci would come to New Mexico to work on the missile autopilot. Ray told me that John Moore had hired Walt Evans and that he was starting a servo course at the plant. Being interested in servos, I asked Ray to try to get me transferred to the servo unit. Walt was just starting his course when I arrived in Downey. *As compared to my earlier servo courses, with a lot of mathematical theory, I felt that Walt had, and was conveying to the students, a true physical understanding.*

The state of the art at that time was that *Transients in Linear Systems* by Gardner and Barnes provided a solid basis for determining a system's behavior using Laplace transforms. In most cases, the open-loop Laplace transform could be derived in factored form without too much effort. The problem was that when the servo loop was closed, a new equation in the form of (1+ the open loop transfer function) resulted.

This required factoring of a high-order equation. If the loop gains or any other parameter was changed, a different equation had to be solved. Methods existed for factoring equations of the fourth order, but higher-order equations required extensive trial and error. At Bell Labs, much work had been done on feedback amplifiers. H. Nyquist published a paper in 1932 on "regeneration theory" where he talks about stability in terms of avoiding the −1 point in a polar plot of gain versus frequency.

One text I remember using was Network Analysis and Feedback Amplifier Design by H. W. Bode. The basic idea was that if the "loop gain" became less than one before the phase shift reached 180 degrees, it would not oscillate. Elaborate schemes were developed to obtain the desired closed-loop frequency response and later to predict the transient response for RADAR pulses. At the MIT Radiation Lab, the same methods were being developed for servos to aim guns, point RADAR dishes, etc. All of the work was done using relationships between amplitude and phase in the frequency regime. Working in the frequency regime had the advantage of making it easy to incorporate test data in the analysis and avoided the problem of factoring high-order equations.

Walt was apparently the first to see that the special form of the closed-loop equation allowed it to be investigated and understood without first generating and factoring a high-order equation. The roots of the closed loop lie on a line (locus) starting at the open loop roots and extending away a distance determined by the gain. The locus could be obtained by summing the angles to the locus from the open loop roots.

Unlike gun directors and feedback amplifiers with constant gain that could be tweaked, I was working on synthesizing an autopilot for a missile. Initially, control was provided by jet vanes in the rocket. As speed increased, control was shifted to aerodynamic surfaces. The control surface effectiveness and other aerodynamic forces changed with both Mach number and dynamic pressure.

The mass and center of gravity of the missile changed as fuel burned. There was no tweaking of the autopilot after launch. The only digital computers at this time were large mainframes used for number crunching. My task was to determine an analog autopilot that would be stable throughout the flight. *Using all the classical analysis methods, I was floundering.*

When Walt first started discussing root locus in his class, *I saw a ray of hope*. While the problem was still very difficult, I could at least get a feel for what I was trying. Initially, Walt used a sheet of semitransparent paper to add the angles for a root-locus plot. As more people started to use Walt's method, many ideas for automating the plot were thought of.

One idea was to place small fish in a tank of water. If voltages were applied at points corresponding to the open loop roots, the fish would line up with lines of equal potential to avoid being shocked.

These lines were the lines of constant gain in root locus, and the locus of roots would be perpendicular to the fish. At this time, Walt was having difficulty getting his first root locus paper published. In true Walt Evans humor, he commented, "*These fish smell as bad as the publishers think my paper smells.*" The publishers and their reviewers were apparently looking for some complex mathematical stuff and failed to see how a graphical method giving approximate solutions was of any value.

Another idea was to stretch a large sheet of rubber over a frame. If the rubber was pushed up a fixed distance at each open loop root, then contour lines on the surface would represent constant gain. A simple second-order system was tried by poking the rubber up in two places. Walt exclaimed, "*That looks just like Sally.*" Sally was a very voluptuous young lady working in the department. None of these early ideas worked out, and I was looking for something better than transparent paper to add angles. I went over to the engineering shop and made an "angle adder" out of a circle of plexiglass with a straight arm held on with a small bolt. This worked much better than the transparent paper for determining the locus, but I still had to measure lengths and multiply them to get loop gain. Walt and I were kicking this problem around one day, and we came up with the idea of adding a logarithmic spiral to my angle adder. This worked well and became the first Spirule. I applied for a patent, but the company decided a copyright was more appropriate.

Meanwhile, Walt was having difficulty getting his first paper published. The editors kept wanting him to make it shorter, as they didn't think it very important or that it contributed much to the state of the art.

One day, he told us that he sent them a version stating that something was difficult to explain but that in the interest of keeping the paper short, the readers should just believe him. Word of mouth and Walt's first paper gradually started spreading the advantages of root locus. *In his second semester, John Moore even had me give a couple of lectures on how to use a Spirule.*

After the first Spirule showed its worth, a design was made for a more precise model. A bunch were made for use in the department. After my demonstration in John Moore's servo class, employees of several aerospace companies asked for a set of drawings. Hughes had several hundred made, but the person doing the assembly looked at the assembly drawing and cemented the parts together in that position. Since nothing could rotate, they were worthless. Walt commented that Howard [Hughes] could afford to build another set but to advise him not to try making them out of wood like his Spruce Goose.

As you know, Walt designed a much simpler Spirule to enclose with his book on servos. He asked me if I wanted to be his partner in the venture. I was busy with other things at the time and declined. I will always remember his kind offer and his desire to be fair.

Speaking of his book, I looked for my copy to see if he had included some things we worked on. I found several servo books but not Walt's. No doubt, someone at Rockwell borrowed it. It must have been very useful because it was my only servo book that was not returned. What I was looking for was whether he included things as "a locus for most anything such as the value of a capacitor in a network" or the "pinball method of factoring high-order polynomials."

Walt had a way of naming things. The pinball method was a good description of how the factoring was done. In teaching root locus, he would start with the locus for a simple quadratic. He would add another term and show its influence. He would explain that if the locus only slightly curved, the added term was "in the outfield" and he called the equation "slightly cubic."

His names were not limited to root locus. Harry Newman placed a shroud around the gyroscopes so the airflow would not change as the gimbals moved. Walt named it a "Newman Coffin."

Norm Parker came up with a way of using precision mechanical drives on NAVAN gyro pairs. Instead of using accurate gyro torquers, one gyro would be mechanically driven at velocity plus earth rate for a fixed time interval. In the Jeff Dole: next interval it would be caged and the mechanical drive returned to zero. Meanwhile, the second gyro would be driven at velocity plus earth rate. Walt named this "the Parker Shuffle."

Len Dozier suggested leaving the gyros un-torqued rather than keeping the platform pointed in a fixed azimuth. *Walt named this "the Dozier Wander."* The first platform used air jet torquers. To balance the platform, Walt placed a penny on it and moved the coin until the air jets pulsed at a very slow rate. He called this "the Penny Fix." We had a short track where the platform could be tested under acceleration. He called this "the Short Snorter." He was an excellent teacher and a fine engineer. I feel that my life was enriched both personally and professionally by knowing him.

Jeff Schmidt

* Emphasis added in all instances.

Appendix 6

Historical Spirule Documents (1948–1951)

In January 1950, Evans presented his root-locus method in a New York meeting of the AIEE. In March 1950, three inquiries arrived, each expressing interest in obtaining a Spirule of the sort Evans had described in New York.

- University of Cincinatti
- Naval Post Graduate School
- Cornell University

These three enquiries were the "flood" of orders Evans would later refer to as evidence that demand would fall "far short of 500."

Years later, the Naval Postgraduate School would place the largest single order for Spirules—over 900 units.

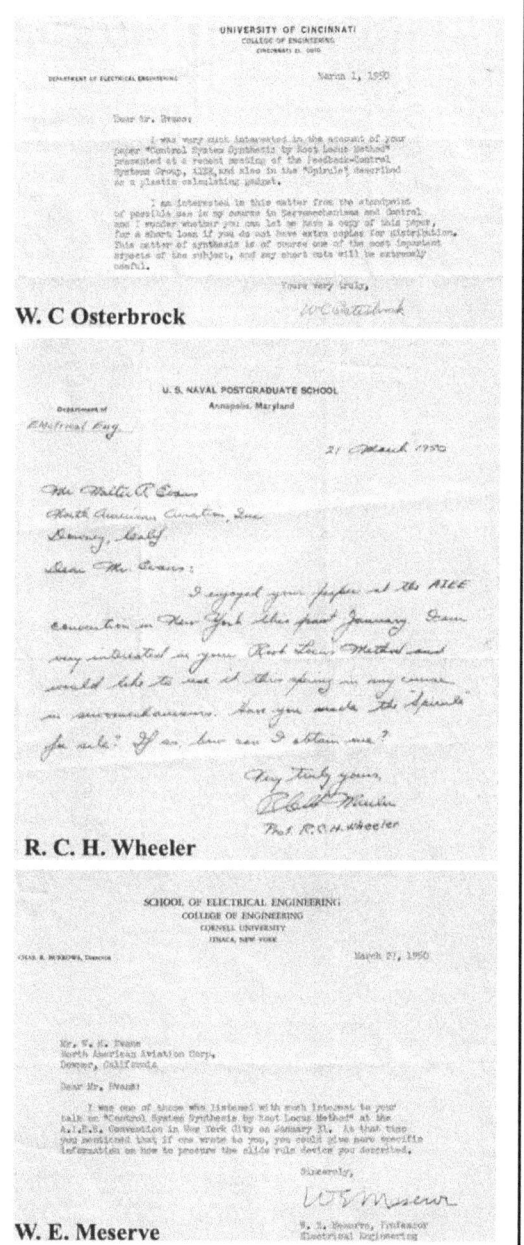

On April 8, 1950, eighteen months before the first Spirule sale and two years before the creation of the Spirule Company, Walt responded to three Spirule inquiries.

In each of these letters, Walt remarked:

"… the following problem in costs: $30 each for a machine shop model, 50c each for a stamped model in a lot of 500. It's not worth $30 and the demand is far short of 500".

On April 8, 1952, Sheldon Saks wrote an inquiry for a Spirule that exhausted the first production run of 500.

The Spirule Company was born.

APPENDICES

W. C. Osterbrock, head of the Department of Electrical Engineering at the University of Cincinnati, responded by telling Evans:

"I wonder whether you are justified in your doubts about the demand for the 'Spirule'. We could have disposed of 65 at a reasonable price, and we shall have another class of about 80 students in servo this summer."

UNIVERSITY OF CINCINNATI
COLLEGE OF ENGINEERING
CINCINNATI 21, OHIO

DEPARTMENT OF ELECTRICAL ENGINEERING

May 3, 1950

Mr. Walter R. Evans
1706 Maple St.
Whittier, California

Dear Mr. Evans:

Thank you for your letter of April 8 and the reprint of your paper. This arrived just at the close of our term so that I was only able to mention your method very briefly to the class.

I have not yet had an opportunity to try out the procedure on an actual problem but am scheduling that for the earliest opportunity.

I wonder whether you are justified in your doubts about the demand for the "spirule". We could have disposed of 65 at a reasonable price, and we shall have another class of about 80 students in servo this summer. Undoubtedly other schools teaching the subject would be glad to recommend it to their students, if properly circularized.

Thanks again for your help, and I hope you will find it possible to have the "spirule" produced and made available.

Sincerely yours,

W. C. Osterbrock
Head, Department of
Electrical Engineering

mjs

At the end of 1951 and the beginning of 1952, two professors with the University of California—Otto Smith of UC Berkeley and Joe Beggs of UCLA—arranged for the university bookstores to place orders for Spirules, thereby enabling servo students to buy from their bookstore.

Over the years, sales to unviersity bookstores would dominate Spirule sales. UCLA's bookstore would buy more Spirules than any other school.

UNIVERSITY OF CALIFORNIA

COLLEGE OF ENGINEERING
DIVISION OF ELECTRICAL ENGINEERING
BERKELEY 4, CALIFORNIA

December 20, 1951

Mr. Walter R. Evans, Specialist
Guidance Section
North American Aviation, Inc.
Downey, California

Dear Mr. Evans:

I would appreciate very much receiving from you copies of your articles on root locus methods of designing servomechanisms. We have a graduate course in servomechanisms in which techniques of this sort are taught, and I would very much like to include your methods and procedures.

If you do not have extra copies of some reports or reprints of some of your papers, I should like the privilege of borrowing them from you for a short time.

Please give my best regards to Dr. John R. Moore.

Sincerely yours,

Otto J. M. Smith
Associate Professor of
Electrical Engineering

OJMS:am

1706 Maple St.
Whittier, Cal.
Jan. 4, 1952

Professor Otto J.M. Smith
Associate Professor, Electrical Engineering
University of California
Berkeley, 4, Cal.

Dear Professor Smith:

I am sending you a copy of the notes which John Moore uses in his class for the root locus method: "The Locus of Roots versus Loop Gain". These notes are actually Chapter 7 of a book which I am now preparing for McGraw-Hill. A copy of the AIEE paper will also be enclosed although the usual reaction to this paper is that it is too short. A short set of notes will also show a quick method of setting up the characteristic equation so that a circuit parameter such as the condenser in a lead network becomes the variable for the locus.

You will note that the Spirule discussed in the notes is not essential to the Root Locus method, approximate plots can be sketched rather easily. A group of them were found however to supply the needs at North American and -- to make a long story short-- I wound up distributing them myself at $2.00 each.

I hope that you find the method of interest to your class and I would be glad to answer any questions which arise regarding its use

Sincerely yours,

Walter R. Evans

APPENDICES

Walt paid $280.67 for an initial manufacturing run of 513 Spirules.

From October 1 through December 31, 1951, Evans sold a grant total of 46 Spirules.

The first college sales were to Caltech's Charles Wilts (2) and MIT's Edward Samario (5).

From these modest beginnings, Arline and Walter kept the letters and purchase orders for every Spirule ordered over the decades to come.

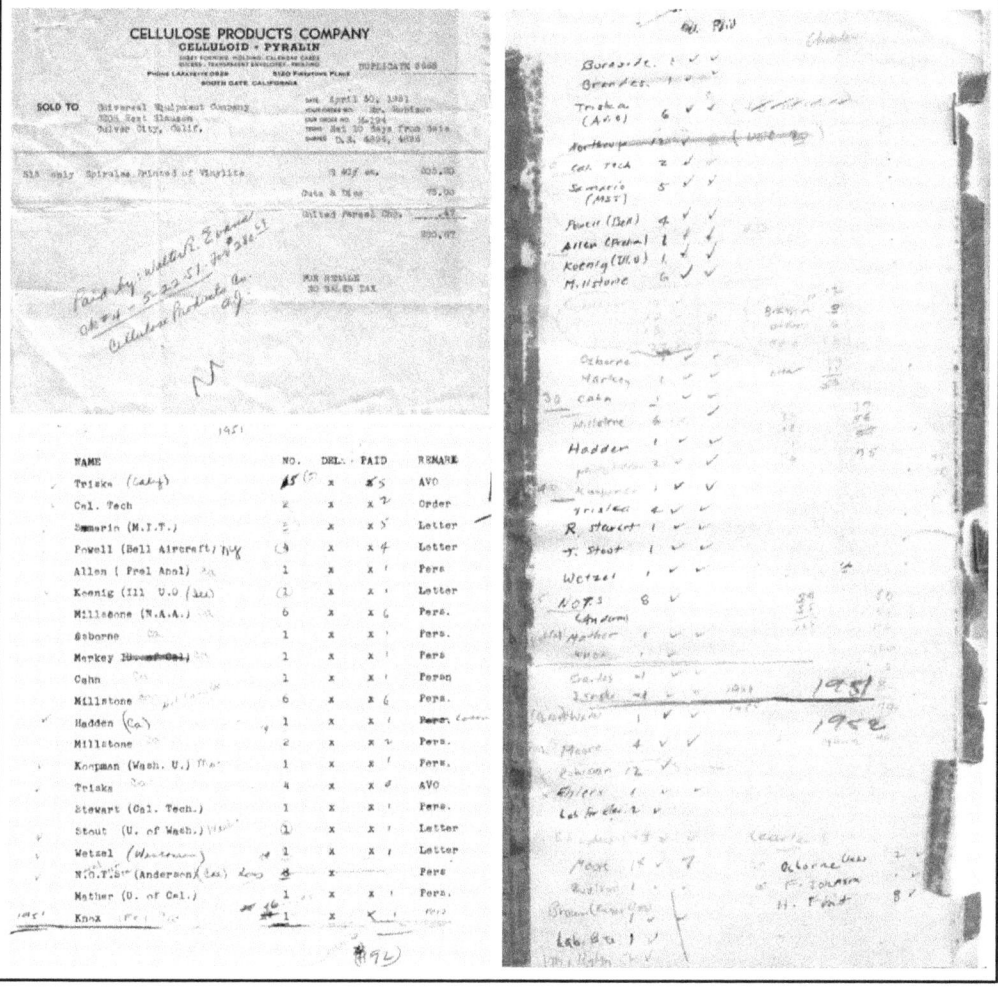

Evans prepared this letter to serve as an interim instruction sheet for the first batch of Spirules.

1706 Maple St.
Whittier, Cal.
October 2, 1951

Dear Sir:

Please pardon the form letter, it is being sent to the large number of engineers who have shown sufficient interest in the Root Locus method of studying control systems to write for North American report AL 787 or the AIEE paper 50-11. The report mentions the possibility of a special plastic device for adding angles and multiplying line lengths and the paper shows a crude model. Sufficient interest has been shown to justify production of the "Spirule"; a photograph appears on p. 594 of the Journal of Aeronautical Sciences September, 1951

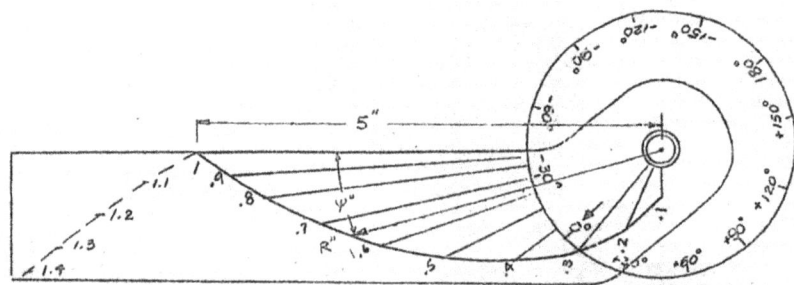

Sketch of the "Spirule"

The disk and arm are held together with a light friction fit by an eyelet; a small pin is held in the center of the eyelet by a clear plastic plug. The pin is stuck in the Root Locus plot at the desired s point to form a pivot for all measurements at that point. Recall that all phase angles and vector lengths on the plot have this one point in common. The total phase angle is obtained by rotating the arm with respect to the disk through each of the angles in succession and is read on the disk at the radial edge of the arm.

The spiral curve on the arm is plotted such that the angle $\psi°$ from the radial edge to the curve is proportional to the logarithm of the radius R to that point on the curve as specified by the equation:

$$\psi°/90° = \log_{10}(R''/5'')$$

The arm is rotated through each of these angles in succession in order to add logarithms. The numerical value of the product is read on the spiral curve in line with an arrow on the disk subject to the correction x^n, in which x is the numerical value on the plot corresponding to 5" and n is the excess of poles over zeros.

Spirules with instruction sheets can be obtained for $2.00 from the undersigned.

Very truly yours,

W. R. Evans
W. R. Evans

APPENDICES

Evans prepared this letter to serve as an interim instruction sheet for the first batch of Spirules.

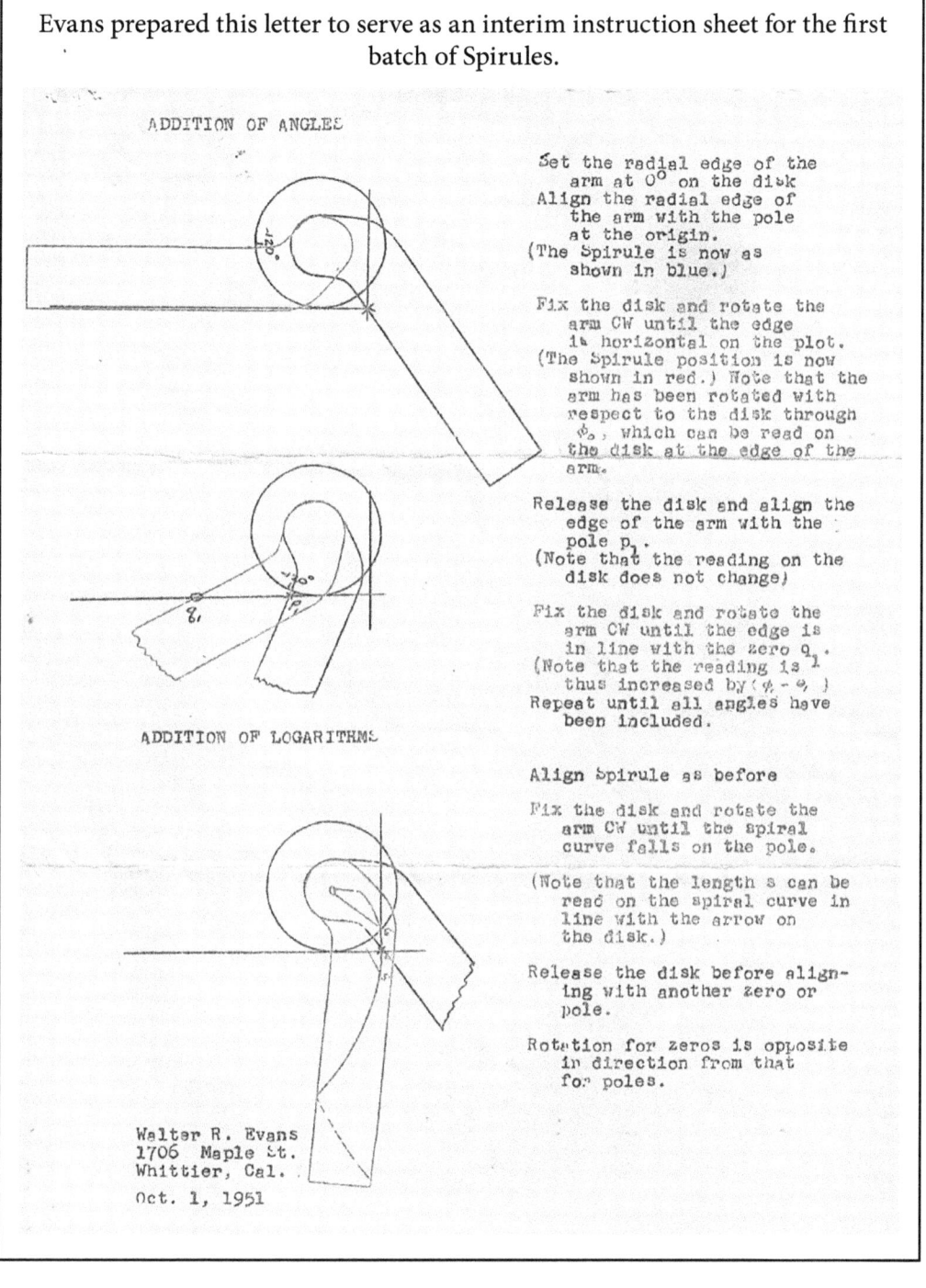

ADDITION OF ANGLES

Set the radial edge of the arm at 0° on the disk
Align the radial edge of the arm with the pole at the origin.
(The Spirule is now as shown in blue.)

Fix the disk and rotate the arm CW until the edge is horizontal on the plot.
(The Spirule position is now shown in red.) Note that the arm has been rotated with respect to the disk through ϕ_o, which can be read on the disk at the edge of the arm.

Release the disk and align the edge of the arm with the pole p_1.
(Note that the reading on the disk does not change)

Fix the disk and rotate the arm CW until the edge is in line with the zero q_1.
(Note that the reading is thus increased by $(\phi_1 - \theta_1)$)
Repeat until all angles have been included.

ADDITION OF LOGARITHMS

Align Spirule as before

Fix the disk and rotate the arm CW until the spiral curve falls on the pole.

(Note that the length s can be read on the spiral curve in line with the arrow on the disk.)

Release the disk before aligning with another zero or pole.

Rotation for zeros is opposite in direction from that for poles.

Walter R. Evans
1706 Maple St.
Whittier, Cal.
Oct. 1, 1951

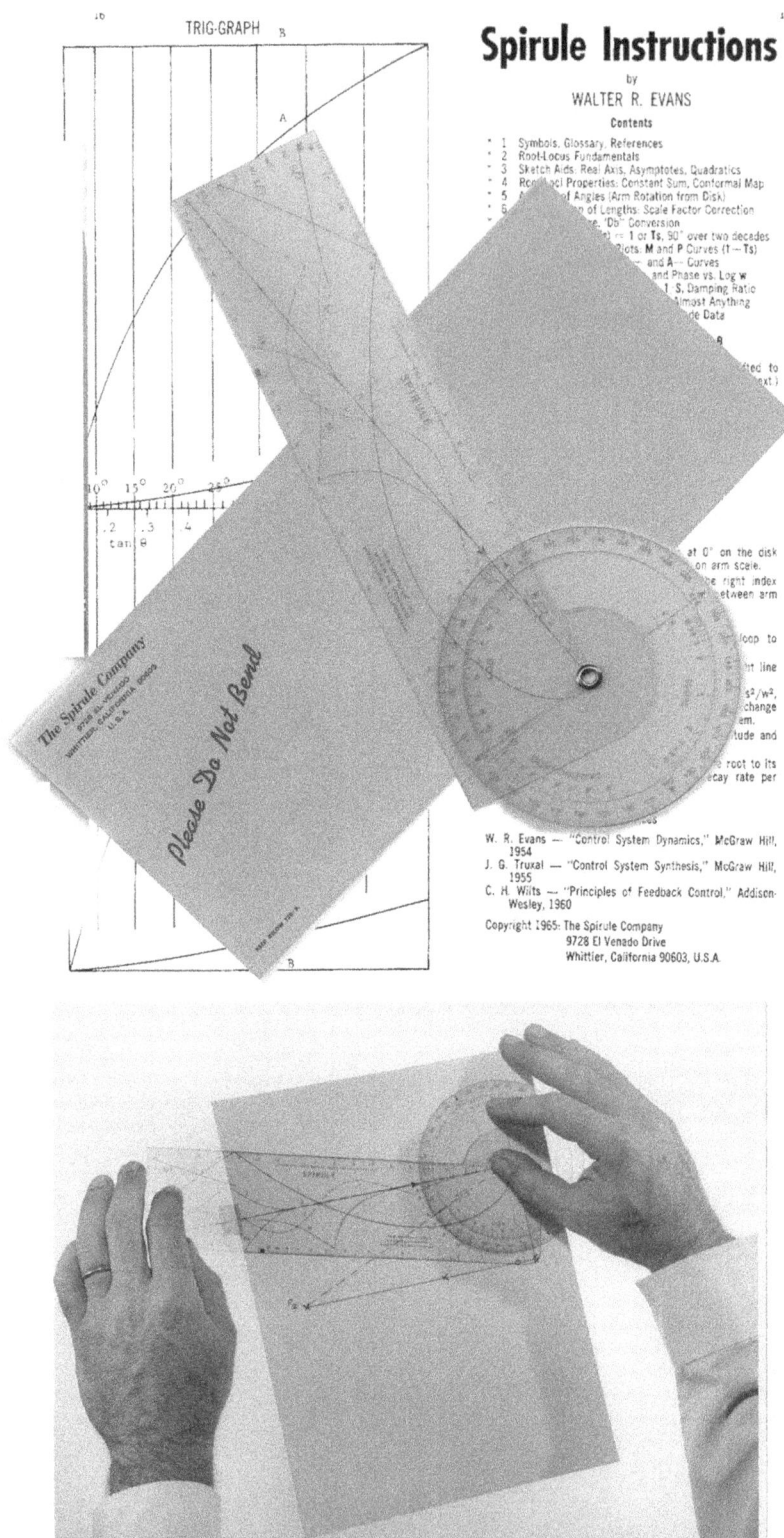

Appendix 7

Profiles of Prominent Engineers

Hendrik Bode (1905–1982) A pioneer in control theory and electrical engineering, Bode developed key principles in feedback control and stability analysis, including Bode plots for frequency response. He worked at Bell Labs and made significant contributions to radar and guidance systems.

William Bollay (1911–1982) was a prominent figure in aerospace engineering. He made significant contributions to the development of jet propulsion systems and was involved in projects that advanced aviation technology. He was nationally known and chief engineer of Aerophysics Lab when Walt joined it in 1948. Walt credited Bollay's 1951 14th Wright Brothers Lecture as a significant contributor to the embrace of root locus and the Spirule by academia and aerospace engineers.

Joe Boltinghouse (1919–2009) was an engineer specializing in the development of inertial navigation systems. Known for his exceptional laboratory skills, he played a pivotal role in advancing navigation technologies, which were crucial for both aviation and maritime applications. Joe was named on twelve patents for gyroscopic design. He was the recipient of the Thomas L. Thurlow Award for outstanding contributions to the science of navigation. Joe and Walt carpooled together to Autonetics in Anaheim in the 1970s.

Gordon Brown (1907–1996) A professor at MIT, Brown was a leader in modern control systems and helped establish MIT's Servomechanisms Laboratory, advancing automatic control in military and industrial systems. Brown's critique of Walt's submissions to the AIEE frustrated Walt. Although his recently published textbook on servomechanisms may have biased him against the paradigm shift root locus represented, he was hardly alone in finding Walt's written descriptions of root locus hard to understand.

Frank W. Bubb Sr. (1892–1961) was a mathematician and scientist at Washington University in St. Louis. Starting as an instructor in Electrical Engineering in 1917, he became head of the Department of Applied Mathematics in 1936. From 1945 until 1949, he headed the Department of Applied Mechanics. Bubb supported Walt's stance on avoiding the Laplace transform in introducing the root-locus method, as few students understood it.

Robert H. Cannon Jr. (1922–2017) A professor at Stanford, Cannon made contributions to aeronautical control, including spacecraft dynamics and underwater vehicle control, influencing modern guidance systems. He wrote the textbook *Dynamics of Physical Systems*. After graduating from MIT, he worked for Walt at NAA. He spoke at Walt's memorial service in 1999 and his Rufus Oldenburger Medal ceremony. He recognized how lucky Walt was to have married to Arline, whom he never failed to address as "Saint Arline."

Tom Curtis was associated with Autonetics. He was instrumental in developing systems for the US Navy's nuclear submarines. Curtis's contributions were crucial during the era when submarines like the USS *Nautilus* (SSN-571) achieved historic milestones, including reaching the North Pole. Curtis accompanied the *Nautilus* on its historic under-ice voyage to the North Pole. Walt always called Tom "Governor."

Charles Draper (1901–1987) Known as "the father of inertial navigation," Draper developed guidance systems for aircraft, missiles, and space exploration. He founded the MIT Instrumentation Laboratory, which played a critical role in Apollo program navigation. He was a mentor of Robert Cannon and an early convert at MIT to the merits of the root-locus method.

Roy Glasgow (1895–1970) Walt's favorite Washington University professor, he was academic dean of US Naval Postgraduate School from 1949 to 1960. Washinton University bestowed him with an honorary doctorate in 1961. Glasgow steered Walt away from an engineering administration degree and encouraged him to enroll in GE's Advanced Course. Walt particularly enjoyed Glasgow's sense of humor. Walt attributed "half my jokes" to Dean Glasgow.

Alexander Suss Langsdorf Sr. (1877–1973) was a distinguished engineer, educator, and author who made significant contributions to engineering education at Washington University in St. Louis. A native of St. Louis, in 1910, he returned to Washington University as dean of the School of Engineering & Architecture, serving until 1920. After a brief period in industry, he resumed his role as dean in 1928, holding the position until 1948.

Orrin W. Livingston (1905–1986), long-time General Electric engineer and inventor, was born and raised in Roselle Park, New Jersey. He began his GE career in 1927 in Schenectady, New York, as a test engineer. At the time of his retirement, he held fifty-seven patents, ranking him among General Electric's leading inventors. He was Walt's supervisor in 1941 and 1942 at GE.

DeWitt Lyon (1922–2015) A contributor to control engineering, Lyon worked on stability and feedback theory, with possible applications in defense and aerospace industries. He left the aerospace industry in the 1950s in order to serve the Lord as a Christian missionary in Japan with his wife and daughters.

John R. Moore (1916–2007) was an engineer and executive who significantly contributed to the fields of inertial navigation and avionics. After earning a B.S. in Mechanical Engineering from Washington University in St. Louis, he led airborne fire control sight and computer development at GE during World War II. In 1948, Moore joined North American Aviation as group leader of their Aerophysics Laboratory Electromechanical Group, was promoted to Director of Engineering for the Autonetics Division, and in 1960 became its president.

Harry Nyquist (1889–1976) A Swedish-American engineer, Nyquist developed the Nyquist criterion for stability in feedback systems, a fundamental result in control theory. His work in telecommunications and signal processing remains highly influential.

Louis T. Rader (1911–2003) was a corporate executive, educator, recipient of the Distinguished Alumni Award from the California Institute of Technology and the Virginia Engineering Foundation Award, and a Fellow of the Institute of Electrical and Electronics Engineers. He was Walt's supervisor at GE in 1945.

Jeff Schmidt (1924–2016) served as a chief engineer at Autonetics in Downey, California, which was renowned for its work in developing advanced navigation and guidance systems. He constructed the first Spirule in 1948 and prepared a patent application, but NAA chose to copyright the device instead.

Fred Terman (1900–1982) A Stanford professor and electrical engineer, his work in feedback amplifiers influenced control theory. In addition, he mentored key figures in engineering and business, playing a key role in shaping Silicon Valley. He was McGraw-Hill's book reviewer for *Control-System Dynamics*.

George Thaler (1928–2018) A professor of control engineering and award-winning teacher, Thaler coauthored influential textbooks on automatic control and worked extensively on military guidance systems. He corresponded with Walt from 1952 to 1982.

John Truxal (1924–2007) A control theorist and educator, Truxal wrote key textbooks on control systems and contributed to signal processing and feedback control applications. He was awarded the Rufus Oldenburger Medal.

Appendix 8

Correspondence from Autonetics Colleagues

1985 Occasion: Official Retirement from Autonetics

Dale McLeod: I have many memories of Walt Evans since I first met him in 1951. To a young engineer then, Walt Evans seemed to me so far ahead of the pack in technical insight and creative ability, as to be regarded only with awe and great respect. Later, as I got to know him better and on a more personal level, my respect for him only increased.

John R. Moore: I understand from Dee Lyon that you retired from Rockwell at the end of January. This is to let you know how much I am thinking of you and remembering all of the great times that we had together, both in St. Louis and California. You must know how much I have always admired your creativity, engineering expertise, and sense of humor, plus your personal warmth and friendship. I think that one of the great desecrations of modern times was the removal of the historic inscription in concrete near the Downey main strip, with its deathless lines that went: *From this point in Euclidean space, Man first attempted to Trace, His path through the heavens, And in spite of Walt Evans, Safely returned to this place.* As I recall, this was the first flight of the XN-1; and an air bump threw you against a power switch on one of the racks, thereby summarily terminating the experiment. You must take a lot of pride and satisfaction from knowing that the root-locus method (and Spirule) has helped many, many thousands of engineers to design better automatic control systems. Walt, please know that I think of you often and you and Arline (with whom you must share my great admiration and affection). I wish the very best for you.

Sam Carlson: It seems only a very short time ago that we met in your office at North American in Downey. That was in September of 1952. Six months earlier I had been interviewed at Purdue University by Ray Hamada, who was on a North American Aviation. I told Ray that if I could work for Walt Evans, I would accept North American's offer. I consider myself most fortunate to have had the opportunity to get started in my career in the Aerospace Industry working for you, Walt. All of us who worked with you learned so much pride in technical excellence, respect for absolute integrity, and satisfaction in accomplishing extremely challenging goals through dedicated individual and team efforts. Walt, I thank you for these lessons. They have been important to me throughout my career and are still today.

Norman F. Parker: Looking back on the days at Autonetics, there was an awful lot of hard work involved but it was enjoyable work because of the people who shared the effort. It was an unusual group, unusually talented but equally important, a team of individuals more interested in the performance of the organization than in their own welfare or progress. I hope when you reminisce you get the same warm feeling that I do with pride in the character and capabilities of your associates as individuals and pride in their accomplishments as a team. Sharing, whether work or humor or support, is an important part of life and you both have given more than your share and God bless you for it.

Bruce Horsfall: I'm glad that De Lyon took the initiative to give some of us a chance to express our best wishes on your retirement. It immediately reminded me of the course you gave a number of us at the NAA Inglewood plant soon after you joined us. Those were days when it seemed that we were getting paid just for having fun. Johnny Moore put together quite a team to develop principles and precision instruments for navigation a couple of orders of magnitude better than previous state of the art, much of which is still fundamental to present equipment, and you were an important team member.

Ray Brands: Just a short note to let you know how much I appreciated working for NAA/Autonetics during the "glory days" of science. But down to business. My admiration for you goes back to the D.C. Machinery class at Washington University in 1947. Your fresh and innovative approach to that subject stimulated my mind and turned what could have been a dull course into one for which I still have fond memories. You were most instrumental in my coming to work at NAA in June of 1951, after I received my MSEE from the University of Illinois. I look back on my decision as one of my better ones.

APPENDICES

1987 Occasion: Rufus Oldenburger Medal from ASME

Jeff Schmidt: In 1948, I was in New Mexico as part of the NATIV missile test crew. Jobs back in Downey didn't seem too plentiful at that particular moment and it looked like my days in the Aerophysics lab would soon be over. Ray Curci used to come to New Mexico every so often to work on the autopilot and he said he thought he could get me into the servo unit. He told me of a new dynamic leader named John Moore that Dr. Edlefson had hired and of a guy named Walt Evans who was teaching a servo course that made sense. This sounded pretty good to me, so I asked him to try and get me a spot. Your course and in particular the root-locus method finally pulled all the parts together for me. I remember the early root locus days when everyone was inventing a way to make plots.

I always thought Walt was one of the best. Many of us can claim a minor role in winning the Cold War but Walt took the black art out of servo analysis. He made it into an understandable procedure for college students worldwide. To top things off he had a great sense of humor.

1999 Occasion: His Passing on July 10

Sam Carlson: Walt Evans was the reason I went to work in North American after graduate work at Purdue University. I had learned of Walt's work in servomechanisms and had read his published papers. When Ray Hamada was recruiting for North American Aviation at Purdue shortly before I completed my graduate studies, I told him I would like to work at North American if I could work with Walt Evans. I reported to work at North American in September 1952, in Walt's group. My immediate supervisor was Walt Pondrom. The two Walts were two brilliant stars. After about six or eight months, Walt Evans launched me on my management career by appointing me supervisor of the Instrumentation Unit, and he was a great person to work for. I have never known anyone more brilliant or of higher personal integrity than Walt Evans.

Frank Pelteson: Walt Evans, of root locus and *Control-System Dynamics* fame, passed away at 79 on July 10. One of his design tools, the Spirule, was used by every servo loop designer before computerization made the problem elementary. Walt interviewed Dr. Sam Carlson and me simultaneously at his Whittier, [California,] home in late 1952 during a football game on TV. During the break he invited us out into the street to lateral some footballs. He asked, during a run-by, "Do you want the job?" I said, "Yes." That was the start of my nineteen-year career at North American. Walt was a simple, straightforward man. He used to disconcert me by solving a problem in five minutes that used to take me four hours to analyze to come to the same conclusion. My immediate boss, Tom Curtis, who was under Walt, on-board serviced and monitored the inertial navigator on the Nautilus North project. The passing of a great genius. And an era, I might add.

Joseph Portney: You described in Walt his genius of seeing the answer or approach immediately where the rest of us falter. This is the unique attribute that many famous scientists possess. Enrico Fermi was said to be able to see the simple approach where others sought out a more complex path that frequently ended nowhere. We live in a great era where both of us have met the great people of our discipline. I was fortunate enough to meet the last of the great navigation practitioners and many aviation pioneers.

2003 Occasion: Support for the Development of This Book

In January 2003, I distributed draft manuscripts of chapters I had written in 2002 to nine of Dad's colleagues. I received responses from John Moore, Robert Cannon, Jeff Schmidt, Hal Engebretson, and DeWitt Lyon. I followed up with interviews in the homes of Robert Cannon and John Moore.

John Moore: I am delighted to hear from you. Your Dad was one of my best friends, and I brought him from GE to Washington U and from Washington U to North American Aviation. I was very familiar with his root-locus method and taught it in my classes from 1950 to 1956 while I was "moonlighting" as UCLA's first Visiting Associate Professor of Engineering in their Graduate School. One of Walt's major characteristics was his sense of humor. In fact, when we were at GE, there came out of the University of North Dakota a song, sung to the tune of "The Battle Hymn of the Republic" called "The Love Life of an Engineer." I added three stanzas to it and your dad added the last two. I am keeping it alive today every chance that I get.

I always had strong affection for Walt and Arline and would like to get to know you and your brother, if that is convenient. ... I look forward to getting together with you.

Best personal regards, John Moore

Hal Engebretson: Walt's support and management style created some of the finest leaders in the Navigation Systems Division (NSD) and Autonetics. Sam Carson became the head of NSD later and went on to manage the start of the Autonetics electronics components business, now mostly nonexistent. George Leisz, another supervisor in Walt's Group 64, followed Sam as head of NSD and later became head of all of Autonetics when John Moore went to the corporate office. As you know, he was a great man.

Jeff Schmidt: Dear Greg, I have been reviewing the Birth of "The Spirule Company." I found it and the GE story both very interesting. My memory of events is a bit different in a few places. Although it does not change the basic story, you might want to consider the following. I would guess that I made my "angle adder" in early September and the logarithmic curve was incorporated by October. I think DeWitt came up with the name Spirule the same week.

Many hand-fabricated "root-locus plotters" WERE NOT MADE. I think I was the only engineer at NAA to make my own. The advantages of the root-locus method and the benefits of a Spirule were enough that within a few weeks a drawing of a much better (and more expensive) model was made and several were built for the key servo engineers. In 1948 a few of us needed Spirules and they were built. I don't know how they were paid for, but they were either charged to the Navaho program or to a John Moore overhead account. ... Fred Rentz prepared Dr. Bollay's Wright Brothers Lecture. Fred was one of the early users of root locus at NAA. He had been a B-17 pilot in WW2 and besides being a good engineer was an excellent "word mechanic."

Stan White: I was first introduced to your dad at a restaurant in Downey in December 1958 when I visited Autonetics on an interview trip.

In 1959 I went to work in the Preliminary Engineering Section of the Inertial Navigation Department. That was where the math whizzes worked out elegant systems-level solutions based on models given them from "on high" where the real engineering was done. The Electromechanical Systems Section was where the real action was ongoing and your dad was Assistant

Section Chief. I transferred ASAP and stayed on the steepest learning curve of my life while working with your dad until I returned to Purdue to finish my Ph.D.

Your dad wrote me a wonderful recommendation and I was able to finish my degree on the NAA Science-Engineering Fellowship program.

DeWitt Lyon: Thank you ever so much for your thoughtfulness in sending me copies of the first phases of your record of your Dad's life and work. What an outstanding job you have done in compiling both the spirit and detail of those years! It is also a tacit testimony to his and your Mom's care in record keeping. I have enjoyed immensely the privilege of going back over some of those years and being able to count him and the family as treasured friends. Not only was there the pleasure of working together but also of returning from time to time from our Japan ministry and being welcomed to "pick up where we left off."

A particular period that stands out to me was the flight testing of the X-1 inertial navigation system. Walt, Jesse Bowman, and I rotated as system engineers with the X-1 in the C-47 (military DC-3) flying out of Downey. Each one would work two shifts, flying with the system in the day followed up by working up the data at night.

To the best of our knowledge at the time, that was the first flying inertial-navigation system anywhere. You probably have lots of information on that period, including the incident when unbeknownst to us a parachute backpack (Walt's as I recall) inadvertently tripped a B+ switch in the gyro power supply. The stabilized platform started to drift, but feedback through synchro's made the scopes monitoring the gyros look as if they were still synched in. Puzzling!

Of particular interest to me is the detail regarding Dr. R. E. Doherty. He was president of Carnegie Tech in the years I was there (1940–43). Dr. Doherty's engineering philosophy which you describe so well was much in evidence at Carnegie. My greatest engineering influence there was from my professor, Dr. B. R. Teare Jr. His methods and courses in Engineering Analysis always started with naming the primary principle to be applied to the problem and proceeding from there. Again, thank you for your diligence given to your dad's biography—a story that certainly needs to be told.

Appendix 9

Walter R. Evans Biographical Information

Walter Richard Evans was born January 15, 1920, in St. Louis, Missouri.

Education
Walt received his B.S. in Electrical Engineering from Washington University in St. Louis in 1941 and his M.S. in Electrical Engineering from UCLA in 1951.

1941–1946: Walt was with the Advanced Engineering Training Program of the General Electric Company in Schenectady, New York, where he taught courses for two years after completing the three-year program.

Employment
1946–1948: He was on the Electrical Engineering staff at Washington University in St. Louis.

1948–1959: He was employed at Aerophysics Lab (renamed Autonetics in 1955, a Division of North American Aviation, now integrated into Boeing).

Walt was with the Inertial Autonavigator Department in charge of the laboratory and first flight tests of the first purely inertial navigator in 1950.

As Systems Group Leader of the Electro-Mechanical Engineering Department, he was responsible for the stable platform and housings for navigator systems.

1957–1959: He was Assistant Section Chief of the Component Engineering Inertial Navigator Engineering Department.

1959–1971: Walt was on the technical staff of the Guidance and Control Department of the Re-Entry Systems Operation of Ford Aeronutronic at Newport Beach, California.

1971–1980: Walt was on the technical staff of the manager of the Strategic System Division of Autonetics, a division of Rockwell International, and served as a "troubleshooter." His advice was utilized in a wide range of programs related to servo systems at many of the Southern California divisions of the company.

Publications

1. "Graphical Analysis of Control Systems," *Trans AIEE*. Vol. 67, pp. 547–551, 1948.
2. "Utilization of Hectograph as a Teaching Aid," paper presented June 14, 1948, ASEE Annual Meeting, Austin, Texas.
3. "Control System Synthesis by the Root-Locus Method," *Trans AIEE*. Vol. 69, pp. 66–69, 1950.
4. "Application of Root-Locus Method to Multi-coupled Systems," thesis, University of California, Los Angeles, 1951.
5. *Control-System Dynamics*, McGraw-Hill Book Company, 1954.
6. "The Use of Zeros and Poles for Frequency Response or Transient Response," *Trans ASME*. Vol. 76, No. 6, 1954.
7. "Vibration Data Needed for Stable Platform Shock Mount Design," 24th Shock and Vibration Symposium, San Francisco, November 1956.
8. "Autopilot Signals Free of Bending Modes Despite Structural Uncertainties," W. R. Evans and Joseph Jerger, Jr., American Rocket Society, Guidance, Control, and Navigation Conference, August 1961.
9. (Contributor) *Inertial Navigation, Analysis, and Design*, edited by C. F. O'Donnell (Chapter III: Shock Mounting Inertial Platform), McGraw-Hill, 1962.

Professional Memberships
ASME, Issued 1964
AIEE Committee on Feedback Control Systems in 1951, 1953, and 1957

Professional Engineer
Licensed as a Professional Engineer in Control System Engineering by the State Board of Registration in California in April 1979

Patents

Patent No. 2,996,631   Filed Jan. 12, 1956; issued 1961
Spring Rate Compensator, Walter Evans

Patent No. 3,139,57   Filed Oct. 23, 1958; issued 1961
Means for Increasing the Accuracy of Synchros, Walter Evans

Patent No. 3,155,437   Filed May 21, 1963; issued 1964
Electromagnetic Bearing W. R. Evans, Eugene Kinsey, R. Brandes, and Bruce Sawyer

Recognitions

1987: Rufus Oldenburger Medal, American Society of Mechanical Engineers (ASME), Dynamic Systems and Control Division

1988: Richard E. Bellman Control Heritage Award, American Automatic Control Conference (AACC)

1990: Engineering Alumni Achievement Award, McKelvey School of Engineering at Washington University in St. Louis

Endowed Scholarship

An engineering student at Washington University in St. Louis is selected to receive an endowed Walter R. Evans Memorial Scholarship.

Appendix 10

The Quotable Walter R. Evans

Walter R. Evans's lifelong passion for geometry played a pivotal role in the development of the root-locus method. His fascination with pictorial thinking began in his Soldan High School geometry class, where he discovered a formal discipline that aligned perfectly with his visual approach to problem-solving. He reflected on his experience in a letter to his son's math teacher:

> *Math has always been a game for me and now is a good part of my livelihood. Geometry used to provide a steady diet of looking for a pattern that would lead to a solution before settling down to the detail of writing down all the steps.*

In a letter to Alexander S. Langsdorf, the former Dean of the Engineering School, Evans reflected on his approach to learning and the impact of his professors:

> *I personally learn most effectively by starting from simple examples and working up. Washington University was excellent in that professors such as yourself, Professor Glasgow, Dr. Bubb, and Dr. Middlemiss could and did take a student all the way back to the beginning if necessary and work up to the question at hand.*

His approach to problem-solving—starting from fundamental principles, applying geometric reasoning, and developing approximate solutions before delving into mathematical detail—became the foundation of the root-locus method. At the time, the concept was unconventional and did not gain immediate acceptance. Reflecting on its success decades later, he remarked:

> *I have had more than my share of luck in hitting it big with root locus. I mean luck because many other ideas, not directly servo, have failed to arouse any interest. I think the explanation is that I strive for a kind of*

understanding that most people don't seek. In the case of root locus, it provided a needed link to a complete solution of a system which many others did seek.

Evans often found himself ahead of his time, proposing ideas such as painting tennis balls bright yellow and orange for visibility, mounting wheels on luggage for easier transport, and staggering work hours to reduce traffic congestion—concepts that later became standard. He was also an early advocate of using computers in the classroom.

In the early 1960s, he and his son Greg built an algebra teaching computer for a science fair project. The machine allowed students to learn at their own pace, adjusting questions based on answers and providing immediate feedback. His primary focus, however, was not on the electrical or mechanical design, but on the structure and content of the questions—how they could best respond to a student's level of understanding. His children sometimes resisted his rigorous approach to learning, preferring more superficial explanations. He once wrote to Dean Langsdorf:

I find that working with my children is a good testing ground for teaching methods because the subject matter is simple, the opportunities frequent, and the reaction clear.

Despite occasional resistance, all four of his children pursued degrees in physics, mathematics, computer science, and engineering, a testament to his influence as a role model. His philosophy toward education was best encapsulated in a 1965 letter to his former professor, Roy Glasgow:

It seems that the real bulk of learning takes place in self-study and problem solving with a lot of positive feedback around that loop. The function of the teacher is to pressure the lazy, inspire the bored, deflate the cocky, encourage the timid, detect and correct individual flaws, and broaden the viewpoint of all. This function looks like that of a coach using the whole gamut of psychology to get each new class of rookies off of the bench and into the game.

Appendix 11

Pinball: Polynomial Factoring with Root Locus

Author's Note: This explanation of the Pinball Method is based upon a 2001 description by Professor Robert H. Cannon Jr. in response to a request from Don Bently, founder of the Bently Corporation. Bently was a former colleague of Evans and credited the success of his company to his knowledge of the root-locus method. ChatGPT took Cannon's original fifth order polynomial, applied the Pinball Method, and computed the root loci depicted in this appendix.

[Dear Mr. Bently,] It is a pleasure to send you an example of the pinball method.

There was a delightful period in my career when I shared Walt's office with him and Bill Mullins. They were two very different, very smart and creative gents. Walt already had in his small office—besides his desk—a large ordinary table with a large paper pad on it (which was known as "the genius pad"). There were always smart people on the other side of the table from him, and a lively technical argument in progress. One day Walt decided he would enjoy having me and Bill in the same office with him. We both jumped at the chance; and Walt had two more desks moved in. We were all solving problems all day, often with consultation from each other. Consultation from Walt was like nothing else.

The easy way for me to tell you about the method I remember is to use it to solve a simple, typical problem. You will then find it easy to generalize use of the method to any polynomial whose roots you would like to know.

But you do have to put it in historical context: There were at that time no digital computers. At all. The Spirule is an ingenious analog substitute with which we were designing control systems full bore at that time (e.g., for inertial navigation stable platforms). [See Appendix 5 for Invention of the Spirule.]

Suppose you have the polynomial. $s^5 + 3s^4 + 7s^3 + 10s^2 + 18s + 80 = 0$

I've used s because (a) I'm used to working with polynomials that are the characteristic equations of dynamic systems, derived by assuming a solution to its differential equations of the form x = C e^(st); and (b) because it takes only one stroke to make an s.

You simply rewrite it in the form:

s (s (s (s (s+3) + 7) + 10) + 18) + 80 = 0

Then you factor each bracketed term in turn, starting with the inside one, and using the Root-Locus Method for each successive factoring, as we'll now demonstrate. (I did all the plots in this treatise with a Spirule! The accuracy is about two and a half significant figures. Today you would have a computer program to plot them for you in seconds with however many significant figures you wanted. In real life your data are typically good to only two or three significant figures anyway; but plotting with a Spirule can take a little time.)

Step 1.

Factoring the first bracket is trivial; but let's plot its locus of roots anyway:

s(s + 3) = -7   Roots: s = -1.5 ± j 2.18

And then check the result algebraically:

$(s + 1.5 \pm j\sqrt{7 - 1.5^2})(s + 1.5 \mp j\sqrt{7 - 1.5^2}) = s^2 + 3s + 7$

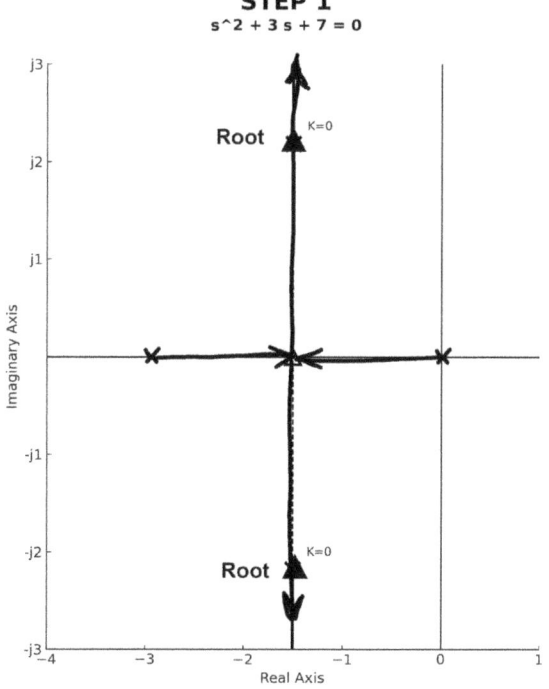

Step 2.

Factoring the second bracket is a classical third-order locus plot. The roots are found to be:

$s = -2, s = -0.5 \pm j\, 2.18$

which, of course, means that the second bracket in factored form is:

$(s + 2)(s^2 + 0.5 \pm j\, 2.18)$

Again, it's easy to check by multiplying the factors out:

$(s + 2)(s^2 + s + 5.0) = s^3 + 3s^2 + 7.0s + 10$

STEP 2

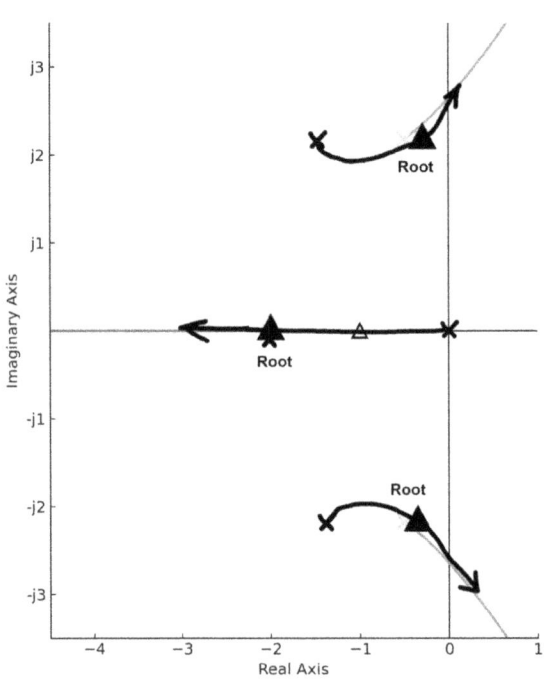

Step 3.

Factoring the third bracket is a fourth-order locus plot, which is a little more complex. The resulting set of roots are:

$s = -1.75 \pm j\, 1.45 \quad s = 0.2 \pm -j\, 1.85$

So that the third bracket in factored form is:

$(s^2 + 3.5s + 5.3)(s^2 - .4s + 3.46)$

When I multiply these factors out I begin to see a bit of error creeping in:

$s^4 + 3.1s^3 + 7.4s^2 + 9.8s + 18.3$

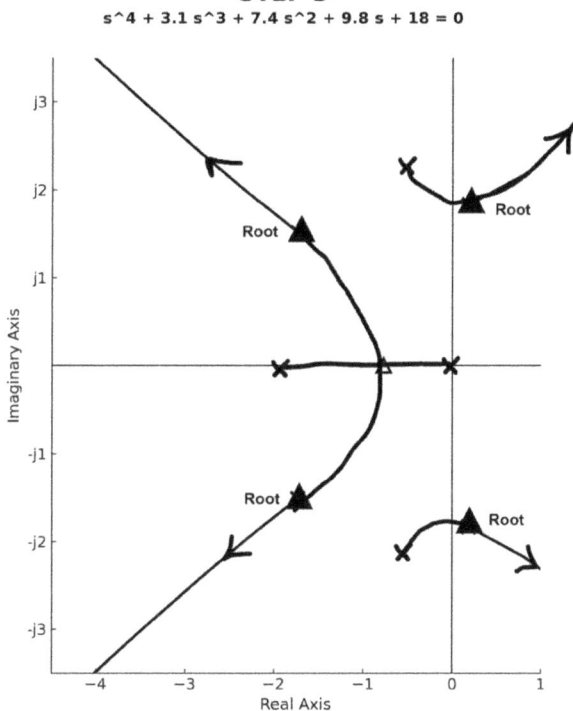

STEP 3
$s\wedge 4 + 3.1\, s\wedge 3 + 7.4\, s\wedge 2 + 9.8\, s + 18 = 0$

INTO STABILITY

Step 4.

Factoring the fourth bracket equation is shown in the figure labeled Step 4. This is a fifth-order root-locus plot. The resulting roots are (to about two and one-half significant figures):

$$s \approx -2.66, \quad s \approx -1.3 \pm j\,2.5, \quad s \approx 1.1 \pm j\,1.7$$

As a check, we simply multiply these factors out to obtain:

$$s^5 + 3.06s^4 + 6.8s^3 + 9.8s^2 + 15.9s + 80$$

This concludes the demonstration of the pinball method, in which each bracket is factored in turn, root loci plotted at each stage, and factors multiplied to verify the polynomial. While originally done with the Spirule, the method translates directly to modern computational tools.

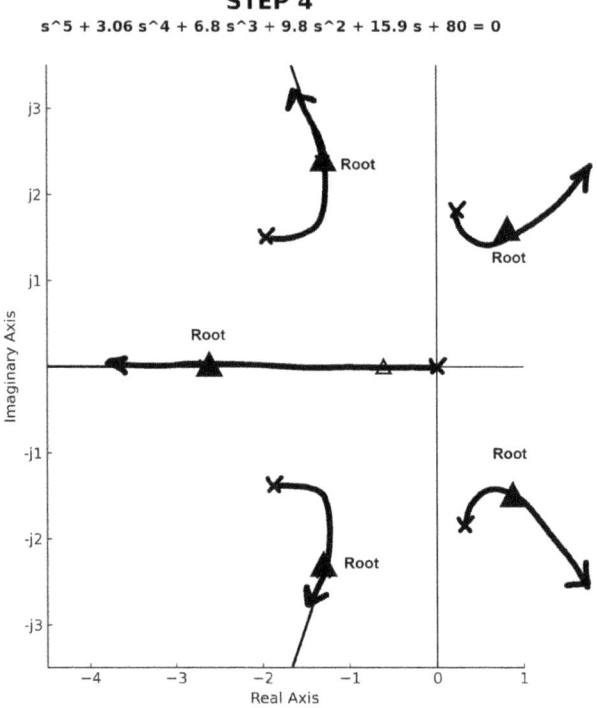

STEP 4
s^5 + 3.06 s^4 + 6.8 s^3 + 9.8 s^2 + 15.9 s + 80 = 0

APPENDICES

Cardinal by Walter R. Evans, Undated, Portfolio #335, Vol. 1

Appendix 12

The Artwork of Walter R. Evans

Walt's artwork resides in a five-volume portfolio of 258 works.

All images are scanned and may be viewed at www.gregevans.zenfolio.com

Walter R. Evans at Senior Care in Whittier, CA. March 1993

Evans Farm Cabin by Walter R. Evans, April 14, 1998, Portfolio #166, Vol. 3

APPENDICES

-109.jpg

-110.jpg

-115.jpg

-117.jpg

-122.jpg

-123.jpg

INTO STABILITY

Walt drew some of his drawings from stamps. Here are nine examples.

These nine drawings were based on stamps.

INTO STABILITY

Cardinal by Walter R. Evans, December 1, 1992, Portfolio #150, Vol. 2

My Stories of
DAD

"Thinking about our four children, it seems to me that the real bulk of learning takes place in self study and problem solving with a lot of positive feedback around that loop. The function of the teacher is to pressure the lazy, inspire the bored, deflate the cocky, encourage the timid, detect and correct individual flaws, and broaden the viewpoint of all. This function looks like that of a coach using the whole gamut of psychology to get each new class of rookies off the bench and into the game."

Walter R. Evans
1965 letter to his favorite professor, Roy Glasgow

Evans men stand at attention
Randy, Greg, Gary, Walter

January 1, 1960 Rose Parade
from our one-night stand

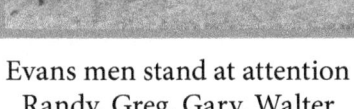

Spirule Company
Home Office

Mimi and Poppy (1995)

With Grandson
Tim Littrell

Nancy and Gary in the
Black Streak (1958)

"Check your move."
With Grandson Tom Evans (1994)

MY STORIES OF DAD

Five Generations of Evans Engineers

When I was born on August 16, 1947, I was given the name Gregory Walter Evans. I am proud to have "Walter" as a middle name. I'm proud too that my son Stephen, a fifth-consecutive-generation Evans engineer, also has Walter as his middle name. Like his grandfather, he earned his undergraduate degree from Washington University in St. Louis.

Walter's grandson, Stephen Walter Evans, entering Washington University in St. Louis in 2010 (left) and graduating in 2015 with a B.S. in Mechanical Engineering (right).

Although I have no memories from the 1940s, I point out that the spark for root locus was Paul Profos's June 1946 journal article. Dad mulled its ideas over in his brain during the nine months I was in Mom's womb. Dad mailed his first letter to Profos on August 11, five days before I was born. Like Dad, I am a son of St. Louis. No matter my age, I am as old as root locus.

Our East Whittier Neighborhood

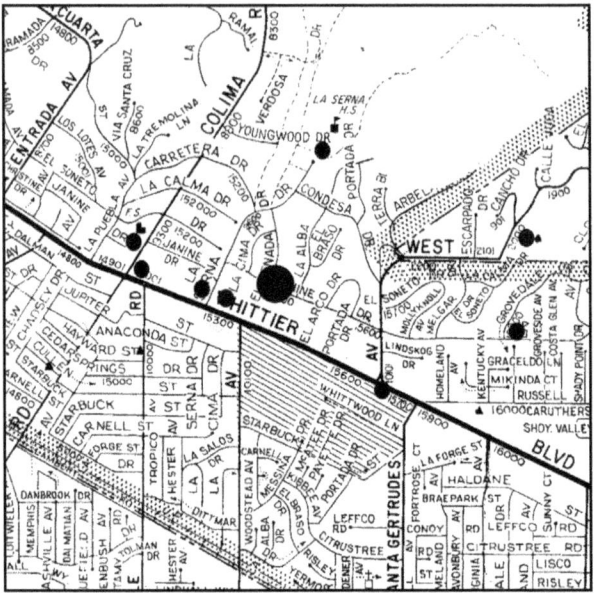

Our neighborhood. Most destinations were within a mile of home (center dot).

Our home since 1954 was located such that everywhere I needed to get was about five minutes away on my bicycle. The figure shows a 2 mile x 2 mile map whose center is 9728 El Venado. Other dots on the map include my elementary school, high school, tennis courts, barbershop, Little League field, grocery store, Whittwood Mall's movie theater, Sav-on drugstore, post office, hardware store, and church. Most are within a half-mile home. Moreover, we lived as far away from a freeway as one could in Los Angeles. The only occasions we had to get on a freeway was to go west into LA for an event (thirty minutes), east to Lake Gregory (two hours), or south to Disneyland twenty minutes). We went to the beach on Beach Boulevard—no freeway. In short, it was a kid-friendly neighborhood in which to grow up.

The Shopsmith

I was in fifth grade or so when Dad bought the Shopsmith. For those unfamiliar, a Shopsmith is the Swiss Army knife of power tools. With remarkable versatility, it could transform from one function to another with relative ease. One moment, it was a table saw; the next, a drill press, then a bandsaw or even a lathe. Its accessories supported ripping and cross-cutting, and its design allowed for both vertical and horizontal operations.

One of its most useful features was its variable-speed motor, controlled by a dial ranging from A (the slowest) to Z (the fastest). It even came with a laminated card specifying the optimal speeds for various materials and operations. Of course, there was also a user manual packed with helpful how-to's and safety guidance on every page. But, heck, Dad was an engineer—he could figure it all out, right? Right. And he did. Safety measures, however, were largely ignored. I count myself lucky to still have ten fingers and two eyes.

Speed control knob for the Shopsmith.

I distinctly remember the very first woodworking project we tackled with the Shopsmith: a four-drawer desk. It featured a top drawer and three side drawers, the bottom-most deep enough for file folders. Every visible surface was covered with a $\frac{1}{16}$-inch veneer of hardwood, finished with two or three coats of clear varnish. The desk measured about 42 inches wide, 30 inches deep, and 32 inches high. And it was mine for the next ten years.

This wasn't a simple project, not by any stretch. But, unless my memory fails me, we started on Saturday morning and completed it by Sunday night, all in that first weekend the Shopsmith graced our garage. I can still recall the hum of the motor, the scent of sawdust filling the air, and the sense of triumph as we attached the last drawer pull. I was a kid, barely ten years old, and I had just helped build a real piece of furniture from scratch. The pride I felt was overwhelming. That desk wasn't just a functional piece of furniture; it was a badge of honor, a tangible reminder of what Dad and I had accomplished together. Countless hours and projects followed with the Shopsmith, but none were as memorable as that desk we first built. It was the first, and it set the tone for everything that came after.

My 1955 Brownie and Dad's 1959 Yashica 8-T2 Movie Camera

My passion for photography started in 1955 at age eight. I took pictures with an inexpensive Kodak Brownie. Once my photos were developed, I entered them into a photo album. I documented every photo in a detailed table of contents. Mom wasn't the only one in our family to keep records!

One of my "trick" pictures: Randy Evans and small person on basketball, backyard (c. 1959).

I liked to take "trick" photos like the one that suggests a small person standing on a basketball. We set up a dark room in the garage for a few months for black-and-white development. I can still recall the odor of the chemicals. Whew!

In 1959 Dad bought a Yashica 8-T2 8-mm movie camera. He allowed me to use it whenever I wanted. It had three lenses and was capable of recording at frame rates from 8 to 64 fps, affording me the ability to shoot 2x speedup or 4x slow motion when played back at 16 fps on the movie projector.

Boy, did I have a good time with that camera—documenting family outings as well as my 1961 experiences at Stanford Coaching Camp and my 1964 summer trip to Germany. Dad's gift endures. Photography became my favorite hobby and it continues to be. It began with encouragement from Dad.

1958 to 1961 Cardinal–Dodger Baseball Games

Like Dad, I was a St. Louis Cardinals fan. When the Los Angeles Dodgers relocated for the 1958 season, it afforded us the opportunity to attend games. Until Dodger Stadium opened in 1962, the Dodgers played in the Coliseum. Remarkably, every August from 1958 through 1961, the Cardinals played the Dodgers in Los Angeles. This took place when I was between the ages of eleven and fifteen, playing catcher in Little League Baseball and more interested in the game than I had been before or since.

We went to a daytime double-header on August 17, 1958, a game on my birthday in 1959, on August 19 in 1960, and August 14 in 1961. I saw my boyhood hero, Stan "the Man" Musial, take his famous crouch in the batter's box.

When the New York Yankees came for an exhibition game on Roy Campanella Day on May 17, 1959, Dad and I joined more than 90,000 fans for the night game. We had nosebleed seats in the top row in straightaway center field, about a quarter-mile from home plate. We followed the action as much with our ears listening to Vin Scully on our transistor radio as with our eyes.

At one point, they turned off the lights and asked everyone to light a candle or a cigarette lighter or a flashlight. Ninety thousand lights filled the stadium. Well, then, from our vantage point, we had the best seats in the place. And no one remembers the game, just that candlelit moment.

Stan "The Man" Musial, c. 1959, Los Angeles Memorial Coliseum

January 1, 1960: Our Viewing Platform for the 1960 Rose Parade

I well remember how I spent New Year's Eve on December 31, 1959. It was in our garage at 9728 El Venado. Mom and Dad were going to take all four of us to the Tournament of Rose Parade on Colorado Avenue in Pasadena on New Year's Day—the first day of the 1960s. I was twelve years old at the time. Dad had gone out and bought a load of 8-foot-long 2 x 4s and loaded them into our 1955 Buick station wagon, along with a generous supply of bolts and wingnuts. We had the radio on in the garage as we listened to midnight celebrations in New York at 9 p.m., Chicago at 10 p.m., Denver at 11 p.m., and Los Angeles at midnight. We cut and drilled the lumber and built a dolly with wheels and a rope to pull it. Before going to bed, we loaded the car with our creation.

Arriving the next morning south of Colorado Boulevard, we piled all the pre-drilled and cut 2 x 4s on the dolly and attached the rope; then Dad pulled the whole assembly (pictured) to our chosen viewing spot east of Hill Street about two miles from the start of the five-mile-long parade route. There we erected a platform that provided a view over the heads of those who had come earlier. The whole family then enjoyed the parade without straining to look over the heads of people in front of us. One envious spectator

Watching Tournament of Roses Parade in Pasadena, CA (Jan 1, 1960)

offered Dad $10 to allow him to watch with us. Dad declined. That's how I spent the first day of the tumultuous 1960s. I still savor the memory.

Dad's Spontaneity and Unselfconscious Acts

Another memory comes from a Cub Scout pack meeting in the multipurpose room of Murphy Ranch Elementary School. Someone on stage asked for a volunteer to join him. I can't remember why a volunteer was needed, but I do remember Dad volunteering and his manner of joining others already on stage. He promptly rose, jogged toward the elevated stage, planted his left hand on the edge, swung his feet up, cleared the edge, and landed on his feet. At the time, I was embarrassed, but I came to admire Dad's unselfconscious approach to life.

1961 Eighth Grade Science Fair Project: An Algebra Teaching Machine

I needed an idea for an eighth-grade science fair project. Dad suggested building a computer that would enable students to learn algebra at their own pace, advancing them to harder problems if they got an answer right and providing a helpful explanation depending upon their wrong answer. In 1961, computers were as big as refrigerators. I wrote algebra questions on a roll of shelf paper, punched binary hole patterns under each answer, and provided the appropriate counts to a stepping relay and direction to a motor drive. The motor engaged when the student closed a sliding wooden panel that triggered a microswitch.

Building digital logic circuits proved to be a lot of fun. Ten years later, I built them for a living at Autonetics. I received my only patent in my career for the design of the digital character generator used in the Metro-Set, the world's first phototypesetter to use fiber-optics. It replaced hot-lead linotype machines.

To my disappointment, Dad was only secondarily interested in the details of the electrical and mechanical design. His primary interests were the contents and structure of its questions and answers: how they could most effectively respond to any student, whatever the state of their initial understanding of algebra.

I learned more than algebra from the teaching machine experience. I learned that writing computer programs and wiring switching relays together were easier and more fun than trying in vain to maintain near-perfect alignment and uniform tension between two rotating wooden spindles configured to move a roll of shelf paper forward and backward. I became an electrical rather than a mechanical engineer.

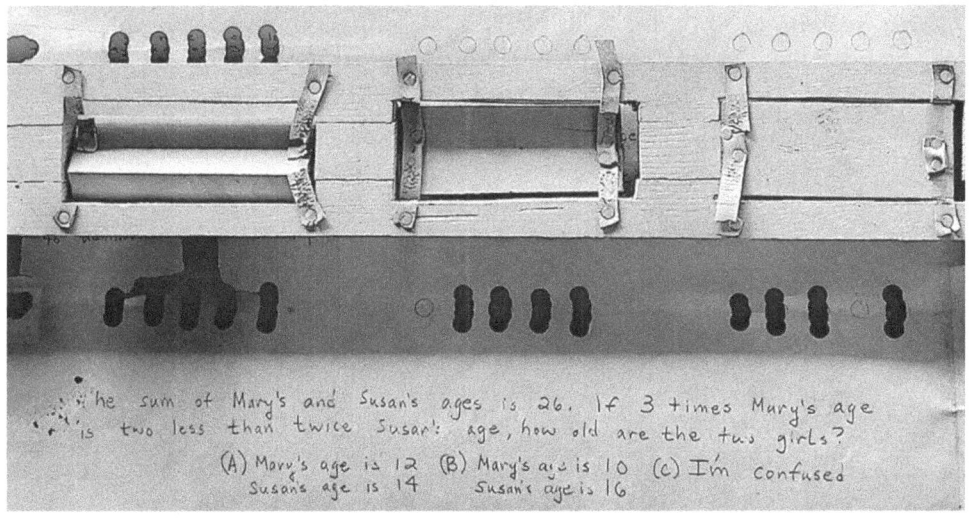

The sum of Mary's and Susan's ages is 26. If 3 times Mary's age is two less than twice Susan's age, how old are the two girls?

A. Mary is 12,   Susan is 14   B. Mary is 10, Susan is 16   C. I'm confused

Weekends with Dad in the Backyard

Shooting hoops

Dad transformed our backyard at 9728 El Venado into a family playground. His own boyhood home at 7048 Nashville Avenue stood beside a vacant lot and had afforded him room to explore. Home movies depicted him riding an Irish Mail down Nashville. And so when Mom voiced a complaint about the condition of our backyard lawn, his response was immediate and unequivocal, "Honey, we're raising kids, not a lawn."

The cable ride was the most notorious, as we had to run along the same patch of lawn to launch ourselves. Dad moved the cable to a location less offensive to Mom when we built the "Black Streak" airplane rides.

Dad was active. Mom remembers her high school principal telling Walter to slow down. It's easy to imagine him running from class to class in the hallways. His upper body was strong. As a youth, he tied a rope to the balcony outside his second-story bedroom on Nashville Avenue and climbed up and down

The "Black Streak" cable ride in the backyard on El Venado (c. 1960)

Dad climbing a rope at Sam Evans's home in Houston, Texas (c. 1960)

Dad riding the cable and wearing a path across the lawn. "Honey, we're raising kids, not a lawn."

it. Years later, in his early forties, when at five-foot-ten his weight was up to about 180 pounds, his brother Sam filmed him climbing a fifteen-foot rope hanging outside his Houston home.

Every summer in the early 1960s, we set up a twenty-four-foot above-ground swimming pool set up in our backyard. Its rim was four feet off the ground. Dad had no problem running straight for the pool, launching himself, and diving into the water from the lawn.

In the winter, we packed inner tubes into the back of the Buick and headed up to the mountains. Dad would drive around Crestline looking for a place we could convert into an inner tube run with a little effort. We took turns taking a ride down. Dad was no exception. He would barrel down the hill with reckless abandon, fly over the snowbank we'd created at the bottom of the hill, and land in the street.

Dad's method of entry into our four-foot high backyard swimming pool was unorthodox. He ran across the lawn and launched himself. (c. 1960)

Inner tubing in San Bernadino Mountains, Crestline, CA (c. 1960)

Discovering Caltech Physicist Richard Feynman

When I attended Caltech, Dad took great interest in Richard Feynman's *Lectures on Physics*, finding Feynman's emphasis on the physical picture and fundamental principles far more illuminating than the physics he had learned. Dad's enthusiasm was contagious, and I switched majors from engineering to physics. Even after Dad had his stroke, his interest in Feynman

INTO STABILITY

Richard Feynman

continued. Mom read him the books by Ralph Leighton (*What Do You Care What Other People Think?* and *Surely You're Joking, Mr. Feynman!*).

When the Rogers Commission issued its report on the cause of the *Challenger* explosion, Mom read Feynman's analysis to Dad. It was politically incorrect for William Rogers, Chair of the Commission, and was relegated to the appendices. Feynman's final sentence is classic: *"For a successful technology, reality must take precedence over public relations, for nature cannot be fooled."*

Learning Root Locus in College Classical Control Theory Classes

I learned about root locus in 1975 from Stanford Professor Gene Franklin, an award-winning teacher. After the final exam, I told him who my father was. He knew Dad and may have interviewed him in July 1959 when Dad visited Stanford at the invitation of Bob Cannon. Anyway, you can imagine his surprise.

Now for the punchline: My younger brother, Gary, took Professor Franklin's class the following year and, like me, waited until the end of the term to introduce himself. Franklin was good-natured. I imagine his reaction must have been something like, "How many more of you are there?"

When my son Stephen learned the root-locus method from Shalom Ruben at the University of Colorado in Boulder, he, too, waited until the end of the term before sharing with his professor that his grandfather was Walter Evans.

Dad Could Defuse Tension with Humor

And always, the sense of humor. Dad defused tense situations with humor. I think he shared this ability with our greatest president, Abraham Lincoln, who was famous for defusing tense situations with a story appropriate for the occasion.

In Dad's case, it would be a joke the punchline of which was appropriate. He had hundreds of punchlines to draw from. He would reconstruct the setup to the punchline on the spot. For example, the punchlines might be *"Now I remember where I left my pencil!"* or *"Now I remember where I left my bicycle!"* In the first instance, the setup might be a nurse asking a doctor why he had a rectal thermometer in his pocket protector. In the second case, it might be a preacher giving a sermon on the Ten Commandments and interrupting himself when he got to the commandment that thou shalt not commit adultery.

Walter Evans c. 1946 in St. Louis, MO

His sense of humor was what first attracted Mom to him in high school and what many of his colleagues loved about him—even repeating favorite quips they still remembered years later. When Dad and a colleague took an early lunch hour in order to beat the lunch hour crowd at a popular restaurant, only to encounter others with the same idea, Dad quipped, *"It's getting harder and harder to beat the cheaters."*

Mimi (Mom) and Poppy (Dad) and My Children

Every year for fifteen years, we packed the minivan on December 26 and drove 350 miles south via I-5 to celebrate a second Christmas with Mimi and Poppy at 9728 El Venado, where Part III began in 1954.

Mom (Mimi) would prepare fresh waffles in the morning and read *Brown Bear* at night. Dad would sing carols while Mom played the piano. We might play a game of Probe, with Dad joining in. On sunny days like the one pictured from 1994, Dad would play chess with a grandchild. And me? I would videotape for posterity, just as I had been doing since Dad loaned me the Yashica 8-mm camera in 1959.

9728 El Venado was home for forty-five years. **Now that's stability.**

Dad playing chess with grandson Tom (c. 1992)

A Surprising January 2024 Email from Athens, Greece

In January 2024, I received this email message from Athens, Greece:

Dear Dr. Evans, We are a group of students from the Physics Department of the National and Kapodistrian University of Athens, who all followed the postgraduate course "Introduction to Control Systems." Together with our lecturer, Dr. Dimitrios Razis, we decided to write this message to you, just to express our admiration for the work of your father, Walter R. Evans.

During the lectures we became truly excited by the concept of the root locus, which we approached following your father's footsteps—constructing the big picture by studying simpler examples. At the end of the semester, in our attempt to share our enthusiasm about these ideas, we decided to create a group—consisting of us along with our lecturer—under the name "Root-Locus." Our shared passion for your father's work has brought us together and [we] have become close friends ever since. Respectfully,

Anagnostopoulos Panagiotis, Anagnostopoulos Theodoros, Bouzalas Dimitrios, Giannopoulos Christos, Uijterwaa Alexandros, and Dr. Razis Dimitrios

ROOT LOCUS: 75 YEARS OLD AND STILL GOING STRONG!

Bob Cannon, Greg, Walter, Arline, Gary
Rufus Oldenburger Medal, 1987 Boston

Front: Arline, Walter, Nacy; Rear: Alice, Betty, Sam
Distinguished Engineering Alumni Award, 1990 St. Louis

Gary, Randy, Greg and Nancy, c. 2003 Anaheim

Cardinals by Walter R. Evans, January 12 1984, Portfolio #110, Vol. 2

Glossary

Bode Plot — A graphical method for analyzing the frequency response of a system, showing gain (magnitude) and phase shift versus frequency.

Classical Control — Techniques like root locus, Bode plots, and Nyquist diagrams used primarily for single-input, single-output (SISO) systems.

Closed Loop — A control system configuration in which feedback is used to automatically correct the system's output.

Control System — A system designed to regulate or manage the behavior of another system using control loops.

Feedback — The process of routing a portion of the system's output back to its input to enhance performance, stability, or accuracy.

Holes — In control systems theory, singularities or undefined points in a transfer function where the function is not finite. These occur when a zero and pole cancel each other out, leaving a gap in the system's response.

Integrate — A mathematical operation that accumulates a function over time, commonly used in controllers to eliminate steady-state error.

Laplace Transform — A mathematical tool used to transform differential equations into algebraic equations for easier system analysis in the s-domain.

Locus — A graphical representation of system pole and zero locations as a parameter (often gain) is varied.

Loop Gain — The product of gains around a closed feedback loop, crucial in determining system stability and response.

MATLAB — A programming environment widely used by engineers, with toolboxes for control design, simulation, and analysis.

INTO STABILITY

Modern Control Theory — A mathematical framework using matrices and differential equations to model and control complex, multivariable systems. Includes state space methods.

Navigation — The process of determining and controlling the position and movement of a system, often using control algorithms.

Nyquist Stability Criterion — A frequency domain method for assessing system stability by examining the Nyquist plot of the open-loop transfer function.

Open Loop — A control system configuration without feedback, where the system operates based only on predefined inputs.

Poles — The values of the input variable that make the denominator of a transfer function equal to zero. They determine system behavior by dictating natural frequencies and stability, with poles in the right-half-plane indicating instability.

Robust Control — Ensures that a system performs well even when there are modeling errors or unexpected disturbances. Focused on worst-case scenarios.

Root Locus — A graphical method to see how the roots of a system (which determine its stability) move as system parameters change.

Roots — The solutions of a characteristic equation that define the system's response, including poles and zeros.

Servomechanism — A closed-loop control system designed to accurately follow or maintain a desired position, speed, or other measurable output. It uses feedback to minimize error by continuously adjusting the input based on the difference between the desired and actual output.

Spirule — A graphical tool used in early control system analysis for determining root locus behavior.

Stability — The ability of a system to return to equilibrium after a disturbance, analyzed using pole locations and stability criteria.

State Space — A way to model a system by tracking its internal state variables. Useful for systems with many interacting parts.

Transfer Function — A mathematical representation of the relationship between a system's input and output in the Laplace domain.

Bibliography

1	Loose Sheet	"The Jump in the Advance of Automatic Control Enabled by Evans' Root-locus Method and His Spirule," by Robert H. Cannon Jr., Unpublished paper for *Into Stability*, 2004
2	Loose Sheets	"The Root-locus Method of Walter Evans," R. H. Cannon Jr., Unpublished paper for *Into Stability*, 2004
3	Talk	Introduction at the Rufus Oldenburger Medal Ceremony Robert H. Cannon Jr., December 1987
4	Article	"Tribute to Walter R. Evans," Robert H. Cannon Jr, "Dynamic System Control Newsletter," Summer 2000
5	Essay	History of Control Systems (ChatGPT)
6	Family History	"His Truth Is Marching On," Greg Evans, Unpublished, 2015
7	Family History	"St. Louis Saints," Greg Evans, Unpublished, 2020
8	Transcript	Washington University Transcript, Walter R. Evans, 1937-1941
9	Article	"The Empire State at War: World War II," New York State War Council, Kurt Drew Hartzell, Ph.D., 1949
10	Obituary	Obituary for Gordon Walter (June 25, 1919 – July 22, 2022), Online @ DignityMemorial.com
11	Interview Notes	Author interview of John Moore at Moore's home in April 2003.
12	Loose sheet	Undated loose sheet with Robert Doherty Quotation
13	Article	"New Challenges for Eng'rs," W.R. Evans, *Schenectady Eng'ring Council Bulletin*, Vol.III No 3, Nov 1945
14	Loose sheet	"Love Life of an Engineer," University of South Dakota, Author and Date unknown
15	Article	"Adv.Course in Engineering," by A. R. Stevenson Jr., and Alan Howard, General Electric, Dec 28, 1934
16	Article	"Post Collegiate Engineering Education," K.B. Mc EachronJr., General Electric, April 1946

17	Article	"Reminiscences of the Adv. Engineering Prog.," Dr. R. Doherty, Carnegie Inst. of Technology, May 1942
18	Journal	"Graphical Analysis of Control Systems," Walter R. Evans, *AIEE Transactions* Vol. 7, January 1948
19	Article	"Legacies Left by Former North American Aviation Experts and Executives," J.R. Moore, unabridged 2002
20	Notes	Design of Servo Mechanisms – Course Notes, Walter R. Evans, 1948
21	Senior Paper	Nancy Evans Littrell Senior Paper, California Polytechnic Institute, 1971
22	Report	Aerophysics Lab Report 787 Control System Synthesis by Root-locus Method, W.R. Evans, March 1949
23	Book	"Chapter 2: Inertial Navigation at NAA 1945–1954" by W.D. Lyon, 13 Jan 1955, in *Twenty Years of Inertial Navigation at North American Aviation*, J. M. Slater, Senior Scientist, 1966
24	Oral History	Interview of Mr. J. Lee Atwood by Mr. Robert Collins, NASM Oral History Project, June 20, 1989:
25	Oral History	Interview of Dr. Robert Burnett by Mr. Robert Collins, NASM Oral History Project, June 19, 1989
26	Journal	"Aerodynamic Stability and Automatic Control," Wright Brothers Lecture, Dr. William. Bollay, December 16, 1950
27	Journal	"Control System Synthesis by Root-locus Method," W.R. Evans, *AIEE Transactions*, Vol 9, March 1950
28	Book	*Automatic Control: Classical Linear Theory*, Edited by George J. Thaler, Benchmark Papers, 1974
29	Book	*Control-System Dynamics*, Walter R. Evans, McGraw-Hill, New York, 1954
30	Article	"Bringing Root-locus to the Classroom," Gregory Walter Evans, *Control Systems Magazine*, 2004
31	Book	*The Navaho Missile Project,* James N Gibson, A Schiffer Military History Book, 1996
32	Poetry	UCLA Engineering Student Newspaper, Vol. VIII, No. 7, March 19, 1956
33	Essay	Enduring Legacy of Root-locus (ChatGPT)
34	Essay	Control Theory Continues to Advance in the Age of Computers (ChatGPT)
35	Book	*Automatic Feedback Control System Synthesis,* John G. Truxal, McGraw-Hill, 1955

BIBLIOGRAPHY

Into Stability List of Correspondence by Chapter, Date, Author, and Recipient

Sent Date	From	To	Sent Date	From	To	Sent Date	From	To
Chapter 1			5/9/49	Walter R. Evans	Orrin Livingston	5/4/50	John W. Wight	Walter R. Evans
5/6/64	Walter R. Evans	Mr. J. Edwin Nettall	5/9/49	Walter R. Evans	Phil Michel	9/29/51	Walter R. Evans	Kenneth Zeigler
5/16/65	Walter R. Evans	Roy S. Glasgow	5/9/49	Walter R. Evans	L.T. Rader	10/22/51	Kenneth Zeigler	Walter R. Evans
3/21/61	Walter R. Evans	A.S. Langsdorf	5/10/49	Gordon S. Brown	Walter R. Evans	10/27/51	Walter R. Evans	Kenneth Zeigler
			5/13/49	Walter R. Evans	Gordon S. Brown	11/4/51	Walter R. Evans	Kenneth Zeigler
Chapter 2			5/13/49	Walter R. Evans	L.T. Rader	11/16/51	Walter R. Evans	Kenneth Zeigler
4/1/42	Walter R. Evans	Mother	5/29/49	Walter R. Evans	Edward C. Day	11/16/51	Walter R. Evans	Fred E. Terman
10/3/42	Walter R. Evans	Mother	6/3/49	Orrin Livingston	Walter R. Evans	11/20/51	Kenneth Zeigler	Walter R. Evans
4/11/42	Herbert Stellwagen	Walter Evans	6/8/49	Shirley L. Clark	Walter R. Evans	11/30/51	Kenneth Zeigler	Walter R. Evans
4/11/42	Herbert Stellwagen	Arline Pilisch	6/8/49	Gordon S. Brown	Walter R. Evans	1/22/52	Walter R. Evans	Elias M. Sabbagh
10/3/42	Walter R. Evans	Mother	6/13/49	Walter R. Evans	Orrin Livingston	4/30/52	Walter R. Evans	Kenneth Zeigler
			8/21/49	Walter R. Evans	John W. Wight	6/5/52	Kenneth Zeigler	Walter R. Evans
Chapter 3			10/20/49	John W. Wight	Walter R. Evans	6/24/52	Walter R. Evans	Kenneth Zeigler
4/7/03	Gordon E. Walter	Greg Evans	3/1/50	W.C. Osterbrock	Walter R. Evans	7/1/52	Kenneth Zeigler	Walter R. Evans
6/16/44	Samuel R Evans	Walter R. Evans	4/8/50	Walter R. Evans	W.C. Osterbeck	10/2/52	Kenneth Zeigler	Walter R. Evans
5/5/45	Walter R. Evans	A-Class students	5/3/50	W.C. Osterbrock	Walter R. Evans	11/8/52	Kenneth Zeigler	Walter R. Evans
			4/1/51	John Truxal	Walter R. Evans	2/11/53	Floyd E. Nixon	Walter R. Evans
Chapter 4.			8/4/49	John W. Wight	Walter R. Evans	2/13/53	Walter R. Evans	Floyd E. Nixon
4/7/03	Gordon E. Walter	Greg Evans	8/21/49	Walter R. Evans	John W. Wight	2/19/54	Jeffrey Norton	Mr. Taylor
4/1/42	Walter R. Evans	Mother	10/20/49	John W. Wight	Walter R. Evans	12/7/54	Kenneth Zeigler	Walter R. Evans
5/16/65	Walter R. Evans	Roy S. Glasgow	11/1/49	Walter R. Evans	John W. Wight	12/11/54	Walter R. Evans	Kenneth Zeigler
12/6/46	Walter R. Evans	C.F. Wagner	11/7/49	John W. Wight	Walter R. Evans	1/27/55	Walter R. Evans	H. W. Burrow
7/30/47	Walter R. Evans	Charles S. Rich	4/19/50	John W. Wight	Walter R. Evans	2/17/55	P.A. Perrone	Ken Hoppens
8/11/47	Walter R. Evans	P. Profos	4/27/50	Walter R. Evans	John W. Wight	2/21/55	Frank Bubb	Walter R. Evans
10/17/47	Charles S. Rich	Walter R. Evans	5/4/50	John W. Wight	Walter R. Evans	2/23/55	Walter R. Evans	Kenneth Zeigler
10/21/47	Walter R. Evans	Charles S. Rich	9/29/51	Walter R. Evans	Kenneth Zeigler	6/15/55	Frank C. Wigdahl	Walter R. Evans
10/24/47	Charles S. Rich	Walter R. Evans	10/22/51	Kenneth Zeigler	Walter R. Evans	12/16/55	Walter R. Evans	Kenneth Zeigler
11/14/47	Walter R. Evans	Charles S. Rich	10/27/51	Walter R. Evans	Kenneth Zeigler	12/22/55	Walter R. Evans	Kenneth Zeigler
12/29/47	Charles S. Rich	Walter R. Evans	11/4/51	Walter R. Evans	Kenneth Zeigler	1/3/56	Walter R. Evans	Kenneth Zeigler
4/22/48	Ross Henninger	Walter R. Evans	11/16/51	Walter R. Evans	Kenneth Zeigler	1/17/56	Kenneth Zeigler	Walter R. Evans
4/27/48	Walter R. Evans	G. Ross Henninger	11/16/51	Walter R. Evans	Fred E. Terman	9/25/56	Walter R. Evans	Kenneth Zeigler
			11/20/51	Kenneth Zeigler	Walter R. Evans	5/7/57	Frank C. Wigdahl	Walter R. Evans
Chapter 5			11/30/51	Kenneth Zeigler	Walter R. Evans			
None			4/30/52	Walter R. Evans	Kenneth Zeigler	**Chapter 10**		
Chapter 6			6/5/52	Kenneth Zeigler	Walter R. Evans	9/30/57	Walter R. Evans	Fred Eyestone
5/16/65	Walter R. Evans	Roy Glasgow	6/24/52	Walter R. Evans	Kenneth Zeigler	8/3/59	Walter R. Evans	R. E. Moore
3/24/04	DeWitt Lyon	Greg Evans	7/1/52	Kenneth Zeigler	Walter R. Evans	11/20/03	Jeff Schmidt	Greg Evans
3/1/85	Walter R. Evans	Randy Evans	10/2/52	Kenneth Zeigler	Walter R. Evans	7/11/99	Sam Carlson	Greg Evans
			11/8/52	Kenneth Zeigler	Walter R. Evans	11/3/03	Hal Engebretson	Greg Evans
Chapter 7			2/11/53	Floyd E. Nixon	Walter R. Evans	3/24/04	DeWitt Lyon	Greg Evans
12/14/03	Jeff Schmidt	Greg Evans	2/13/53	Walter R. Evans	Floyd E. Nixon	7/11/99	Robert Nease	Greg Evans
10/1/49	Charles H Witts	Walter R. Evans				3/13/85	John Moore	Greg Evans
1/8/51	J. E Chadwick	Walter R. Evans	**Chapter 9**			2/23/59	Walter Evans	Donald Fisher
2/1/51	Walter R. Evans	Walter R. Evans	8/4/49	John W. Wight	Walter R. Evans			
3/8/52	UC Berkeley	Walter R. Evans	8/21/49	Walter R. Evans	John W. Wight	**Chapter 11**		
			10/20/49	John W. Wight	Walter R. Evans	2/16/59	John G. Truxal	Walter R. Evans
Chapter 8			11/1/49	Walter R. Evans	John W. Wight	7/13/73	Ho Hwa Hui	Walter R. Evans
5/7/49	Walter R. Evans	Gordon S. Brown	11/7/49	John W. Wight	Walter R. Evans			
5/9/49	Walter R. Evans	Gordon S. Brown	4/19/50	John W. Wight	Walter R. Evans	**Bonus**		
5/9/49	Walter R. Evans	Gorden Walter	4/27/50	Walter R. Evans	John W. Wight	1/25/24	Dimitrios Razis	Greg Evans

Acknowledgements

This book was inspired by my father's colleagues at North American Aviation (NAA), whose bonds of affection supported Dad throughout his engineering career.

Their recollections of experiences they shared with him during his transformative ten year period from 1945 to 1954 form the beating heart of *Into Stability*.

I am especially indebted to Gordon Walter, Robert Cannon, Jeff Schmidt, DeWitt Lyon, and John Moore for their extensive written accounts. Norm Parker, former President of Autonetics, contributed photographs of Dad's coworkers.

This book was made possible by my mother's meticulous record-keeping—hundreds of letters, originals of those received and carbons of those sent, carefully preserved and organized. Having access to these documents made writing this book possible; managing the sheer volume of correspondence made it a challenge.

I wish to thank the five engineering professors who provided comments and wrote endorsements based upon draft manuscripts. I am accountable for any and all errors.

My sister, Nancy, who read and offered comments on every draft, deserves credit for the book's keeping in balance its two story lines: Walter Evans and Root Locus.

I am thankful for the patience of my wife, Carol, who endured my hours of research, my takeover of a spare bedroom, and my 24/7 sessions typing at the computer.

Those hours were not wasted. Even if no one ever reads what I've typed, I will always treasure the joy of reliving good times with Dad.

Contributors

Developmental Editor	Robert Weinstein
Copy Editor	Lindsay Evermore
Interior Designer	Catherine Williams
Cover Artist	Jonathan Sainsbury

Foreword Appendix 1 Appendix 11	Robert H Cannon Jr.
Appendix 3	Gordon Walter
Appendix 5	Jeff Schmidt
Before Root Locus Legacy of Root Locus After Root Locus	ChatGPT-aided

Colleagues who corresponded (in alphabetical order)

Ray Brandes	Bob Cannon
Sam Carlson	Hal Engebretson
Bruce Horsfall	DeWitt Lyon
Dale McLeod	John R. Moore
Robert Nease	Norm Parker
Frank Pelteson	Joseph Portney
Jeff Schmidt	Stan White

Photo Credits

The author gratefully acknowledges the following sources for photographs and illustrations included in this book.

University Archives and Libraries
- Fred Terman—Stanford University Archives
- Roy Middlemiss—Washington University Photographic Services Collection, Julian Edison Department of Special Collections, Washington University in St. Louis
- Robert Doherty—Courtesy of Carnegie Mellon University
- George J. Thaler—IEEE History Center (https://ethw.org/George_J_Thaler)

Hatchet Yearbook Portraits (Washington University in St. Louis)
- Alexander Langsdorf—Dean of Engineering, 1941 Hatchet, p. 22 (scan from the author's library)
- John R. Moore—Senior class portrait, 1941 Hatchet, p. 38 (scan from the author's library, Photo-Reflex Mirror-Camera Studios)
- Walter R. Evans—Senior class portrait, 1938 Hatchet, p. 52 (scan from the author's library, Photo-Reflex Mirror-Camera Studios)

Courtesy of Families and Colleagues
- Frank Bubb—Courtesy of John M. Bubb, grandson of Frank W. Bubb, Sr.
- Roy Glasgow—Photograph taken by Walter R. Evans (scan from author's library)
- Jeff Schmidt—Taken at an Autonetics reunion in the early 2000s by family
- DeWitt Lyon—From personal correspondence in the early 2000s

Museums and Historical Collections
- William Bollay—MIT Libraries, Cambridge, Massachusetts, GCP-00002498 (adapted)
- Gordon S. Brown, ca. 1970—MIT Libraries, Cambridge, Massachusetts, GCP-00002945
- Tom Curtis aboard the Nautilus—"Autonetics Guidance System is 'Superb,'" *Downey Skywriter* (Aug. 15, 1958)

CONTRIBUTORS

- XN-1 Gyroscope—J. M. Slater, *Twenty Years of Inertial Navigation at North American Aviation* (North American Aviation, Autonetics Division, 1966), p. 15

Public Domain Images
- Paul Profos—Photo by Wilhelm Pleyer. ETH Library Zurich, Image Archive / Portr_04621
- Archimedes Woodcut—WikiMedia
- Richard Feynman—*The Big T* (1959), yearbook of California Institute of Technology, via WikiMedia
- C-47 Transport Aircraft—Photo by Adrian Pingstone
- General Electric Plant Map—General Electric, *Manual for Employees, General Electric Company Schenectady Works* (1926)
- General Electric Schenectady Works Plant Map—General Electric, *Schenectady Works Welcomes You!* (1949)
- Navaho Missile—Air Force Flight Test Center (AFFTC) History Office
- Aerial View of St. Louis—Courtesy National Parks Service.

Personal Collection of the Author
- John R. Moore's Bullpen—Received from a NAA colleague after 1980.
- Robert Cannon—Photograph at Rufus Oldenburger Medal Ceremony (digital scan)
- North American Aviation—Bullpen photos of individual engineers, 1950 dinner party, spirule blueprint, and 1959 seating chart (digital scans from author's collection)
- Cartoon of C-47 Flight Test—Created by ChatGPT
- Cartoon of bicyclist—Created by ChatGPT
- Map of Schenectady—Google Maps. Map data © 2025 Google.

Photos unlisted above were all obtained from the digital scans of photographs in the author's personal collection. The author went to great lengths to identify copyright holders for photographs listed and published in this book, including the use of the professional research services With Permission. Any additional information regarding sources or rights is warmly welcomed.

About the Author

Gregory Walter Evans was born in St. Louis in August 1947. He has lived in California since John Moore hired his father to work at North American Aviation's Aerophysics Laboratory in 1948.

He received a B.S. in Physics from the California Institute of Technology in 1969 and an M.S. in Electrical Engineering from Stanford University in 1975.**

Carol and Greg Evans, Glacier Bay, Alaska, July 2008

He was a member of technical staff at Rockwell International from 1965 to 1977, a Technical Fellow at TRW (now Northrop Grumman) from 1977 to 2008, and a part-time employee of VTS (now Voyager) from 2008 to 2023.

He is author of "Bringing Root Locus to the Classroom," published in *IEEE Control Systems Magazine* (Vol. 24, Issue 6, December 2004).

Greg maintains two websites: *gregevans.zenfolio.com* for his photography and his father's artwork and *walterrevans.com* for information about his father.

He has three grown children and one grandchild and lives with his wife, Carol, in Los Altos, California. He can be reached by email at greg@walterrevans.com.

** Stanford Dean of Engineering, John Linvill, called the author into his office after he failed the oral Ph.D. qualification exam. Linvill told him that he had the highest score of all the non-qualifiers. What the author knew and Linvill did not know was that the author had failed to answer a question on the root-locus method asked by Professor Bernie Widrow that required only a rudimentary understanding. That failure may well have led to his not qualifying.

www.ingramcontent.com/pod-product-compliance
Lightning Source LLC
Chambersburg PA
CBHW061943130526
44582CB00042B/98